Preface

The object of this book is to offer enthusiast modellers as much *visual* coverage of the Avro Shackleton as possible within the confines of just 96 pages – the term 'visual' not being used lightly as the space thus required precludes any suggestion that this book might present a fully-comprehensive written history of the Shackleton. Much of the feedback received by us following the publication of Flight Craft 3 and 4 suggested that a reduction in body text would be useful, particularly if the space saved could be used for additional well-captioned photographs. The inference, it seems, is that a preference exists for caption text rather than body text. This poses a question: does such feedback represent the view of a silent majority of readers or those of a verbal minority? It's hard to say of course, but either way, such remarks are worthy of consideration based on the adage that if a single picture is worth a thousand words then several additional images ought to be worth several thousand more words! Coincidentally, in reducing the body text we were able to include several appendices to help compensate for the loss of text, while simultaneously allowing us to incorporate about 20% more photographs than in Lancaster. The real difficulty this time being which 160 or so images should be included out of the 330 made available to us by very generous friends and colleagues. We hope you approve our choice and that readers will feel we have more or less achieved a useful balance and, as ever, that occasional modellers might be tempted to buy a copy too.

Acknowledgements

The authors would like to express their grateful thanks to Roger Lindsay, Paul Lucas and Fred Martin for their invaluable contribution, help and assistance throughout, as well as our appreciation to Martin Blundell, Alan Carlaw, Derek Hemingway, Donald MacKay and R.A. Scholefield for allowing the authors to use their Shackleton images which were so gratefully received. Our gratitude is of course, once again, further proffered to the unflagging help and support from our friends Mike Smith, Rosalyn Blackmore and all of the staff at Newark Air Museum who have opened their archive and facilities to us on so many occasions. Finally, always last to be acknowledged it seems, but *never* least, our special thanks are again extended to Mark Gauntlet for his superb artwork.

Notes

Other than a few references to wartime and early post-war aircraft which were assigned Roman Mark numbers, all aircraft Mark numbers in this work appear in Arabic.

For a short period the earliest Shackleton Mk.1s, possibly limited to just the prototypes and the first couple of production machines, were initially referred to as GR (i.e. general reconnaissance) aircraft at a time when a change to MR (maritime reconnaissance) was in vogue. For ease of reference we have only used the designator MR in reference to the Shackleton.

As it was not possible to accurately date every Shackleton photograph obtained, any attempt to present them in chronological order proved impossible to achieve, hence it was felt that a useful alternative would be to arrange them by serial number – commencing afresh within *each* section. This at least makes it relatively simple for readers to search for a specific serial number and check whether a particular aircraft has been included or not. A further benefit arises whereby, if more than one image of a particular aircraft is included, it allows those images to appear close to each other enabling readers to take note of any change in appearance. Inevitably there are exceptions – one of which arises immediately because, for reasons explained later, the three prototypes received serials in the VW range while the first production Shackletons received VP serials. A further exception exists within the camouflage and markings section where images are displayed to illustrate changing colour schemes primarily, rather than the aircraft themselves.

Martin Derry and Neil Robinson

An atmospheric image of an unidentified MR.3 on short finals. *Author's collection*

A Defined Purpose

Sunderland GR.5, NJ267 'A-B' No.201 Squadron. One of the few remaining UK-based Sunderlands, this aircraft porpoised and sank whilst taking off at Pembroke Dock on 3 March 1954. RAF Sunderlands continued to operate from the UK until 1957, but remained in service until mid-1959 in the Far East.
Author's collection

The Avro Shackleton was primarily an instrument of RAF Coastal Command. Outlined by the Air Ministry in 1946, the Command's defined purpose was to analyse, investigate and develop modern methods of enforcing or achieving the defence of friendly maritime trade against all forms of attack other than from the air – a purpose that applied equally to the RAF's maritime reconnaissance units stationed elsewhere in the world. Wedded to this was the requirement to *attack* enemy ships and submarines as opposed to merely reacting to *their* attacks.

In 1946/47, complying with the Air Ministry's requirement was easier said than done given the stringent post-war financial economies which were applied to virtually every aspect of British life, none more so than within the nation's armed forces. Additionally, the end of the war meant that items supplied to the UK under the terms of the wartime Lend-Lease Act had to be returned in the near future, scrapped or paid for. The latter was not an option, so it was fortunate therefore that the threat of general conflict at this time was low.

From Coastal Command's perspective, the end of Lend-Lease meant an imminent return of American-supplied long-range maritime Boeing Fortress' and Consolidated Liberators and Catalinas. This left the Command with a gaping capability gap that couldn't be filled by existing indigenous designs – primarily the Short Sunderland flying boat, and general reconnaissance (GR) variants of the Vickers Warwick (a derivative of the Wellington bomber) which in any case was soon retired. Fortunately (as related in Flight Craft 4 *Avro Lancaster In Military Service 1945-1965*) Britain did at least enjoy a surplus of Avro Lancasters and consequently several Lancaster B.3 airframes were set aside and adapted for maritime patrol duties, the B.3 being selected due to the Mark being fitted with Packard-built Merlin engines. Unsurprisingly the Lancaster proved successful in its new role and remained in service with Coastal Command until as late as 1956, but it was an adaptation – a stop-gap, hence a purpose-built design was urgently required to supplement and later replace both the Lancaster and Sunderland.

To leap ahead of ourselves temporarily, a real sense of urgency crept into proceedings in 1950, when intelligence sources perceived that a large increase in the size of the Soviet Navy's surface fleet was underway, to be matched by a greatly increased submarine fleet – provoking memories of Hitler's U-boat

Avro Lancaster MR.3, SW366 'H-Z' seen in May 1956 wearing the final colour scheme applied to RAF Lancasters. Introduced in 1955, the overall Dark Sea Grey Scheme was accompanied by white serials and codes, although at least three MR.3s received red underwing serials and codes (outlined in white) with plain red serials on the fuselage. *Roger Lindsay*

offensive just scant years earlier. Although the first Shackletons entered RAF service in 1951, the quantities needed to transform Coastal Command's capabilities in coping with the Soviet threat would not become available for a further three years, thus the RAF obtained US-supplied Lockheed P2V-5 Neptune maritime patrol aircraft through the Mutual Defence Assistance Programme in order to help plug the capability gap. The first of fifty-two examples to be received by the RAF arrived in the UK in January 1952.

Operationally, by early 1955, Coastal Command wielded nine Shackleton and two Sunderland squadrons to patrol the eastern reaches of the Atlantic from the UK and Gibraltar, while four squadrons of Neptunes patrolled the North Sea and areas more northerly. (Two further RAF units had re-equipped with Shackletons at Malta in 1953, while Sunderlands soldiered on in the Far East until 1958/59). Unsurprisingly of course, that perennial feature which dogs all

of Britain's defence procurement quickly reasserted itself and so, during 1957, all the remaining Neptunes and (UK-based) Sunderlands were disposed of.

The Shackleton continued to guard Britain's maritime interests until it in turn was replaced by the (then) Hawker Siddeley Nimrod maritime patrol aircraft (MPA). However, as recent history reminds us, Nimrod would not be replaced and today Britain has no fixed-wing MPA whatsoever – an absurd situation for an island nation in which 95% of its imports arrive by sea at a time when the equally underfunded Royal Navy musters a mere thirteen frigates and six destroyers to cover world-wide commitments as well as the forthcoming need to escort Britain's two new aircraft carriers! Consequently, when unlawful submarine activity occurs *inside* British coastal waters, the UK today has to rely upon the goodwill of other nations to supply an MPA to help police *our* sovereign waters!

Sunderlands at bay. *Author's collection*

Above: Lockheed Neptune MR.1 (mod), WX501 '4' which was accepted by the Air Ministry on 20 November 1952. This particular Neptune was somewhat unusual in that it was one of four issued to Fighter Command's Vanguard Flight and not to Coastal Command. Formed on 1 November 1952 at Kinloss, Vanguard Flight began airborne early warning (AEW) trials on 12 January 1953 with three MR.1s on strength. On 5 June 1953 the unit moved to Topcliffe to become No.1453 (Airborne Early Warning) Flight until disbanded on 30 June 1956. In addition to WX501 (sold as scrap in November 1958) the Flight also operated WX499 '2' and WX500 '3' until joined by WX542 after 22 April 1953. Interestingly an unusual link exists between the four Neptunes and the Shackleton. RAF Neptunes were fitted with AN/APS-20, a very capable surface search radar, which, with minor changes, offered a good AEW capability in the hands of well-trained operators. Thus, while the Royal Navy was receiving fifty Douglas Skyraider AEW.1s equipped with APS-20 for its carrier fleet, the RAF was trialling APS-20 in the AEW role too. Skyraiders were replaced by APS-20-equipped Fairey Gannet AEW.3s until they were also withdrawn as the RN's carrier fleet was progressively scrapped. In order to close an ensuing AEW radar gap, two decades after Vanguard Flight first stood up, twelve Shackleton MR.2s – redesignated AEW.2s, received a later variant of the same basic radar. *Author's collection*

Below: 'L' an otherwise unidentified Neptune MR.1 equipped with underwing rocket rails, 20mm nose and (presumably) tail cannon, with twin .50 mgs in the dorsal turret. Fifty-two Neptunes were acquired by the RAF serialed WX493 to WX529 and WX542 to WX556, and although it doesn't always apply, in this instance WX556 was indeed the last Neptune to be accepted (on 15 November 1953). Neptunes served operationally with Nos.36, 203, 210 and 217 Squadrons. *Author's collection*

Shackleton MR.1 Enters Service

On Wednesday 28 May 1947, Avro received a contract to build three prototype Avro Type 696 maritime patrol aircraft which subsequently received the serials VW126, VW131 and VW135. The Type 696 would become the first British land-based, long-range, four-engined aircraft specifically designed for the maritime patrol and reconnaissance role – a role still fulfilled, in 1947, by flying boats and adapted bombers (namely Sunderland GR.5 and Lancaster GR.3).

A year or so earlier, the growing requirement for a new, dedicated, maritime patrol aircraft had been recognised as essential. Because early post-war aircraft procurement could be a protracted affair, it was hoped that a solution might be found quickly, thereby allowing Coastal Command's Lancasters to be replaced in the not too distant future. The solution, it was thought, lay in what was hoped would be a relatively straightforward adaptation of the RAF's existing Lincoln B.2 bomber, a type that in early 1946 was already set to steadily replace Bomber Command's Lancasters.

The new maritime aircraft was soon being referred to as the Lincoln 3 and, possibly because of the pedigree of its forebears, a contract was placed for thirty aircraft in March 1946, virtually off the drawing board. Given that the 'Lincoln 3' incorporated several assemblies from the Avro Tudor airliner as well as the Lincoln, plus a higher-set tailplane with redesigned fins and rudders and a fuselage that was deeper, shorter and wider than the Lincoln, the Type 696 was in fact an entirely new design. Hence the later order for three prototypes. The serial number allocations for the thirty production MR.1s were: VP253-VP268 & VP281-VP294 of which one, VP253, was cancelled when the prototypes were ordered fourteen months later – thus the production machines received serial numbers which, alphanumerically at least, preceded the prototypes.

The first Type 969 to fly was prototype VW126 which made its maiden flight on 9 March 1949 powered by four 12-cylinder Rolls-Royce Griffon piston engines, each driving a pair of contra-rotating propellers. Just three Marks of Griffon were ultimately employed by Shackletons across the years, namely Marks 57, 57A and 58, each of which produced approximately 1960hp at maximum power, with water methanol injection providing another 500hp if required.

With regard to weaponry, confusion has persisted for years concerning how many guns early Shackletons (a name chosen by Avro's chief designer: Roy Chadwick) actually carried. Whereas all production MR.1 and MR.1As were initially fitted with twin 20mm Hispano cannon in a dorsal turret, prototypes VW126 and VW131 differed somewhat. It has been stated that the former received a full complement of guns including a tail turret, although the authors have been unable to locate a photograph proving it. It is true that VW126 was temporarily fitted with gun barbettes, positioned one each on either side of the nose (an arrangement imposed by the positioning of the nose-mounted radar which dictated that the two front guns be separated), it being possible that the barbettes were actually aerodynamic mock-ups.

VW131 also received two nose barbettes, each with a 20mm Hispano placed in a ball mounting within each barbette. Reportedly only one tail turret was ever fitted to a Shackleton, if so then VW131 was the recipient and not VW126 as can be seen in the accompanying photos. The tail turret was a Boulton Paul Type D unit equipped with twin .5in machine guns. Both barbettes and the tail turret were soon removed to leave VW131 with just a dorsal turret.

As related in the introduction, there was a pressing need, from as early as 1950, to get the Shackleton MR.1 into operational service as soon as possible to supplement Coastal Command's Sunderlands and Lancasters as the perceived threat of Soviet naval expansion began to emerge.

The first *operational* unit to have a Shackleton permanently allocated to it was 120 Squadron at Kinloss which received VP260 on 30 March 1951, followed by VP258 and VP259 in early April. Also at Kinloss was 236 Operational Conversion Unit which received its first Shackleton (VP264) on 31 May 1951, the two units undertaking service trials of the new type. By the end of 1951, the OCU had received a dozen MR.1/1As, while 120 Squadron had received eight MR.1s.

Other units receiving Shackletons in 1951 included the Air-Sea Warfare Development Unit (ASWDU) which received VP261 on 27 April 1951 – for just three weeks before it was transferred to 120 Squadron. Number 224 Squadron at Gibraltar received its first Shackleton MR.1 in July, while 220 Squadron reformed in September at Kinloss. (For further details concerning units and dates please refer to the appendices).

Two prototype Shackletons accompanied by a production machine – possibly VP154. The one nearest the camera is VW126, with VW135 in the middle. VW126 first flew on 9 March 1949 while VW135 first flew a year later on 29 March 1950 – one day after the first production Shackleton – dating this image to late March 1950 at the earliest. It would appear that VW126's first year was quite intensive judging by the heavy staining across its wings during which time its nose barbettes had been removed. VW126 had been transferred to the Air Ministry by November 1959 and afterwards received the maintenance serial 7626M; it was broken up in 1965. VW135 was broken up in April 1954, both aircraft having spent their respective lives conducting tests and trials. *Author's collection*

On 1 August 1951, the first Shackleton MR.1A, WB818 made its initial flight, the primary difference between the two variants being a slight widening of the Mk.1A's outer engine nacelles to enable them to accommodate the Griffon 57A engine, thus making inner and outer engines interchangeable. Most surviving Shackleton MR.1s were similarly modified during the middle years of the decade.

Following the demise of Coastal Command's last Lancasters and Sunderlands in 1956 and 1957 respectively, and with the prior introduction of the Shackleton MR.2, early model Shackletons didn't survive much longer either. They too had been largely removed from operational duties by the end of the decade, with just 205 Squadron in the Far East retaining some into 1962. For many aircrew, perhaps the majority, the Mk.1's passing was not mourned, for although it had indeed proved itself to be a particularly good maritime aircraft, for aircrews flying without the benefit of any interior sound-proofing it could be a noisy, uncomfortable and fatiguing experience.

However, not all MR.1/1As were consigned to ground instructional duties or stored pending scrapping. Three MR.1s, VP258, VP259 and VP293 were converted into trainers from 1955 and 1956 and redesignated as T.4, with VP258 effectively becoming the prototype. The need for a Shackleton trainer had been foreseen a year or two earlier, while in 1956, plans were put in progress which would see Coastal Command standardise on the Shackleton for its long-range maritime patrol duties, with Neptunes and UK-based Sunderlands being phased out in the near-term, along with those Lancasters that still remained in use in the aircrew training role. Further, training procedures were to be streamlined and brought under the umbrella of the Maritime Operational Training Unit (MOTU) on 1 October 1956 which would be created by merging 236 OCU and the Maritime Reconnaissance School.

Thus the need for the T.4 became obvious, particularly as it was a crew-trainer which, apart from having dual-controls for its pilots, was also fitted with stations to instruct the other specialist trades that together comprised a Shackleton crew. Included also was ASV 21 radar, an eagerly awaited improvement over the earlier, somewhat unreliable, ASV 13. Of course, three Shackleton trainers would never be sufficient to meet the new training requirements and so fourteen further conversions were authorised, all from MR.1A stocks. (See appendices). Surviving T.4s were retained in service until 1968, the year in which they were replaced by ten converted MR.2s – redesignated as T.2 – with conversion of at least two more being halted before completion. The first T.2 to enter service was WL739 which joined MOTU on 3 January 1968 and, as with the MR.1/T.4, there were few physical external differences by which it might be possible to distinguish an MR.2 from a T.2 unless the former retained nose guns.

VW131 as seen on 10 September 1949 with its rear turret visible. The nose gun installations and the rear turret proved unsuccessful and were soon removed. *Author's collection*

A Company promotional image of their new Shackleton maritime patrol bomber, in this case the third prototype, VW135, as seen from below. *Newark Air Museum*

Demonstrating a capability that the UK no longer enjoys, a Shackleton MR.1 polices the sea lanes. First flown on 18 September 1950, this publicity image of VP256 was taken in 1950/51, predating the aircraft's issue to 224 Squadron on 30 August 1951 with whom it received the code 'B-A'. Operated by 269 Squadron from February 1952, yet retaining the same code, this aircraft was seriously damaged and subsequently dumped when, on 26 October 1954, it failed to become airborne when the elevators, which had not been properly unlocked, jammed on take-off – just one of seven RAF aircraft which crashed and were written-off that day including: two Meteors, one Mosquito, one Jet Provost and two Sabre F.4s. *Author's collection*

The first T.4. Following conversion, VP258 underwent various periods of trials until, in January 1959, it arrived at Kinloss where it was briefly assigned to the Maritime Operational Training Unit (MOTU) coded 'D'. Sent to and fro in a series of extended modifications this T.4 was returned to MOTU for the last time in 1963 when it received the code 'N' seen here. VP258's final allocation was to the Stanstead Fire School where it had arrived by July 1968 and where, presumably, it finally expired. *Author's collection*

MR.1, VP262, was first flown in February 1951 and was delivered to 120 Squadron two months later where it received the code 'A-D'. The bulbous profile of the rear fuselage clearly betrays the fact that a rear turret was originally intended for the type. Transferred to MOTU in October 1956, VP262 was sent into storage in September 1958 and broken up in 1962 or 1963. *Author's collection*

MR.1, VP287, first flew in June 1951 and was issued to 224 Squadron at Gibraltar a month later coded 'B-B'. Transferred to 269 Squadron in February 1952 it remained as 'B-B' (both units used the code 'B'). VP287 was transferred to 240 Squadron on 5 June 1956 with whom it remained until delivered for storage with 23MU, Aldergrove, in October 1958 where it was scrapped five years later. Interestingly, there is no evidence of an individual code letter in this image – only the Squadron number being evident supported by the unit motif on a white disc on the nose. *Author's collection*

MR.1, VP288. First flown on 15 June 1951, by the time this photograph was taken VP288 had participated in a number of trials and served with the ASWDU and 220 Squadron before being placed in storage at 23MU in November 1958. Withdrawn from the latter, this aircraft was allocated to 205 Squadron in the Far East arriving at Singapore in mid-September 1959 coded 'K'. VP288 was sold for scrap in August 1964. *Roger Lindsay*

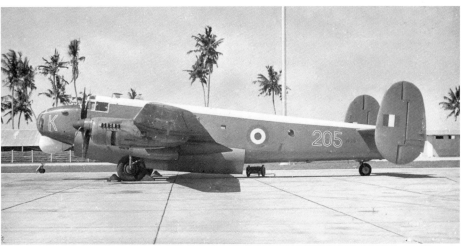

First flown in July 1951, MR.1, VP292 was allocated to 236 OCU from early October 1951 receiving the code 'C-S' and, as can be seen, suffered an accident (in October 1952) when the undercarriage was raised too soon. VP292 went on to serve with MOTU and later with 205 Squadron at Changi, Singapore, from August 1959. It was struck off charge there on 28 April 1961. *Author's collection*

Above: The longest-lived Shackleton MR.1 was VP293 which first flew in July 1951. Allocated to 236 OCU two months later, the aircraft was subsequently operated by a number of squadrons when not undergoing various modifications. Converted to T.4 standard from August 1956, VP293 was later operated by MOTU prior to being stored at 23MU from February 1963. The recipient of several modifications during its life so far, this T.4 was selected for use by the RAE at Farnborough in January 1964 for trials work with a distinctive colour scheme. VP293 was retired on 23 May 1975. *Courtesy of Donald MacKay*

Below: VP293 was later sold to the Strathallen Aircraft Collection where it arrived on 3 May 1976 and where these images were taken later that year – before its unique colour scheme began to weather and fade. They must have been taken at different times however as the port props vary slightly. Put up for sale when the Collection closed in September 1988, VP293 could not be saved and was broken up on site in February 1990. *Fred Martin*

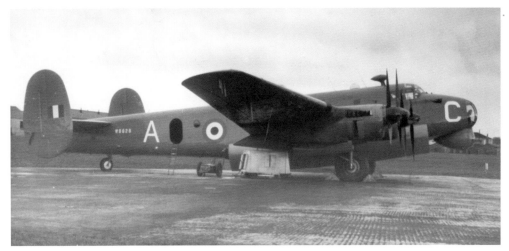

Shackleton MR.1A, WB821 'T-L', from 220 Squadron is seen at St Eval in early 1954. First flown in August 1951 this aircraft arrived with 220 Squadron two months later and remained until transferred to 236 OCU in mid-April 1954 with whom it was coded 'C-Z'. WB821 later went on to serve with 206 Squadron (1956) and MOTU from September 1959 with periods in between for modifications. Withdrawn from service in early 1961, WB821 was scrapped the following year. *Roger Lindsay*

MR.1A WB828, probably seen in September 1951 shortly after its first flight to judge from the lack of exhaust staining. It was allocated to its first operational unit, 220 Squadron, in early January 1952. *Newark Air Museum*

Having served with 220, 204 and 240 Squadrons and undergone various modifications and repairs, WB828 was allocated to 120 Squadron in September 1955 and coded 'A-C' as seen here. It appears that the dorsal turret has been removed, but in fact it is still in situ with just the smallest portion visible above the wing's leading edge. *Author's collection*

Still definitely a member of 120 Squadron, WB828 is probably seen in mid-to-late 1956 prior to being assigned to 206 Squadron in November 1956 with whom it was coded 'J'. This aircraft was subsequently operated by 240 then 204 Squadrons (from mid-1958) until it was sent for storage in early 1960. WB828 was sold for scrap in November 1962. *Author's collection*

MR.1A, WB829 'O' seen after 1 October 1956, the day on which MOTU was formed following the disbanding of 236 OCU and the SMR/MRS. First flown on 21 September 1951, WB829 was assigned to 236 OCU three months later and received the code 'C-O'. Other than a short period in storage, it remained with the OCU until passing into the hands of the newly-formed MOTU still coded 'O'. Damaged in May 1957, WB829 was sent for repairs and modifications which were completed in late 1958, following which it went into storage. Allocated to 205 Squadron in the Far East in August 1959, WB829 was grounded soon after arrival with wing fatigue issues and was struck off charge (SOC) on 28 April 1961. *Newark Air Museum*

Centre: First flown in September 1951, MR.1A, WB831, was issued to 220 Squadron three months later and remained with the unit until February 1956 when WB831 was despatched to Langar for conversion to T.4 standard. In late November 1957 this machine arrived with MOTU and was coded 'S', as seen in this image, which was retained until October 1965 when it became 'U'. WB831 crashed following a roller-landing and take-off at St Mawgan on 17 May 1967 when the undercarriage was raised prematurely causing the starboard propellers to strike the ground putting the aircraft on its belly. WB831 was SOC shortly thereafter and scrapped in 1968. *Author's collection*

Above: T.4, WB837 'W' seen whilst serving with MOTU following its relocation to St Mawgan in 1965, a move which also prompted a change in WB837's code from 'H' to 'W'. Withdrawn in May 1968, seventeen years after its first flight as an MR.1A, in October 1951, WB837 was sold for scrap in February 1969. *Author's collection*

MR.1A WB844 first served with 224 Squadron from January 1952 and remained with the unit until reassigned to 120 Squadron on 30 August 1954 coded 'A-F', and later 'F' once changes to the unit code system had been implemented as indicated here. *Roger Lindsay*

Following service with 120 Squadron, WB844 was converted to T.4 standard in mid-1956 and in January 1958 was allocated to MOTU, coded 'L', which it retained until the unit moved to St Mawgan in 1965 at which point WB844's code was changed to 'R'. The T.4 was flown to Cosford in July 1968 where it received the maintenance serial 8028M and was scrapped there *circa* 1970. The white panel positioned aft of the code letter contains the unit motif, adapted from MOTU's badge, and depicts a seal. *Author's collection*

Delivered as an MR.1A to 224 Squadron in February 1952, coded 'B-O', WB845's conversion to a T.4 began in late 1960 and was completed in January 1962. WB845 'X' is seen whilst serving with MOTU between July 1965 and July 1968 – the month in which it was placed in open storage with 27MU at Shawbury prior to being scrapped in 1969. *Newark Air Museum*

MR.1A, WB846 'B-P' photographed between 5 February 1952, the date this aircraft was issued to Gibraltar-based 224 Squadron, and late August 1954 when it was transferred to 120 Squadron with whom it became 'A-G'. Subsequently transferred to MOTU in October 1956, WB846 was withdrawn from use in March 1958 and received the maintenance serial 7561M at Kinloss where it was ultimately scrapped. The aircraft seen below WB846's nose, and aft of its tail, are Vickers Valetta transports. *Newark Air Museum*

Above: WB847 'Z' was a MOTU aircraft that had been recoded in mid-1966, two years before its final withdrawal in June 1968. Sadly, neither the date nor the location of this photograph was recorded, although there is a suggestion that it was taken at Kinloss, to which location WB847 had been flown prior to an abortive attempt to save it, the hope being that it would become a gate guardian there. *Author's collection*

Below: Instead WB847, seen at Lossiemouth in August 1976, was later dumped and used for fire training but not before acquiring the code 'B'. *Derek Hemingway via Fred Martin*

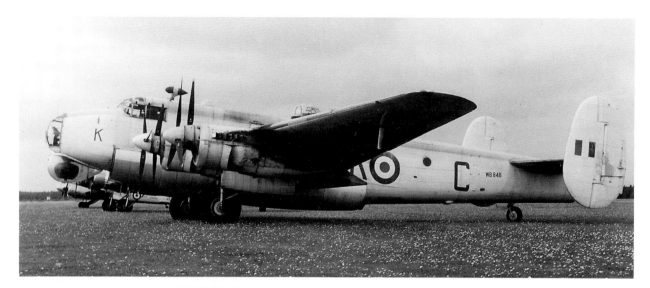

Above: MR.1A, WB848 'C-K', from 236 OCU seen in 1953. Delivered to the OCU in February 1952, it stayed through the unit's transition to MOTU in October 1956 and later served briefly with 240 Squadron in late 1957. Following a rather short operational existence, WB848 was despatched to 23MU in January 1958 and remained there until sold for scrap in October 1963.
Via Roger Lindsay

Centre and above: Port and starboard views of MR.1A, WB850, of 240 Squadron. These images were taken after 240 Squadron had dispensed with their unit code 'L' and the aircraft's transfer to 204 Squadron in June 1958. It served with the latter until December 1958 when it was sent to 23MU where it remained until scrapped in May 1963. WB850 displays the Squadron motif on its nose in a panel containing a winged Icelandic helmet (not to be confused with 228 Squadron which used a similar device) commemorating the time spent by the unit in Iceland in WWII. Below the motif is what appears to be the word *Mjölner* (pronounced myul-nir) – which is plausible given it is the name of Thor's hammer, the Norse god from whom our Thursday is derived! *Roger Lindsay* and *Author's collection* respectively.

MR.1A, WB851 'G-Y', from the Ballykelly-based Joint Anti-Submarine School (JASS) Flight which was formed for the purpose of practising anti-submarine tactics. (Although the three JASS Shackleton MR.1s received individual codes in the W-Y range, the letter displayed looks more like a V than Y!) The Flight was commanded jointly by RN and RAF personnel while its Shackletons were distinctively marked – with twin black bands on their outer wings and a single, broader, stripe around the fuselage near to mid-point. *Author's collection*

First flown on 12 December 1951, WB851 was delivered to JASS Flight in March 1952 where it remained until transferred to 269 Squadron who operated it for a year from September 1954 until September 1955. Thereafter WB851 served with 220, 206, 269 and 204 Squadrons prior to joining ASWDU in November 1959. Placed in storage in early 1960, WB851 was sold for scrap in February 1963. *Author's collection*

The relevance of this image showing MOTU T.4, WB858 'M', relates to the position of the fuselage roundel which in so many instances was invariably masked by the aircraft's wing if photographed from a forward quarter. Although this image is not dated, it is known that WB858 received the code letter 'M' in April 1964 and retained it until the aircraft was withdrawn from use in July 1968. *Newark Air Museum*

Work-stained MR.1A, WG528 'B-E', from 206 Squadron seen after July 1954 when its code was changed from 'B-V' to that shown here. In February 1958, WG528 was sent to 23MU for storage until sold for scrap in May 1963. *Author's collection*

Operation *Grapple* was the name given to a number of tests (*Grapple 1, Grapple X, Grapple Y, Grapple Z*) covering the development and testing of Britain's first hydrogen bombs at and from Christmas Island in the Central Pacific from 1956 to 1958. Shackletons were not involved in the dropping of such weapons, but they were required for air-sea-rescue, reconnaissance and patrol duties to be certain that ships and other vessels remained clear of the test zone. Although several Shackleton units were ultimately involved with *Grapple*, directly or otherwise, this image taken at Christmas Island in 1957 shows MR.1A WB828 'J' from 206 Squadron undergoing an engine change. *Newark Air Museum*

An unidentified MR.1 or MR.1A coded 'V'. Given its early colour scheme, dorsal turret and relative paucity of the use of 'V' as an individual aircraft code amongst Mk.1s, the probability is that this is MR.1A, WG528 of 206 Squadron which became 'B-V' in October 1952, prior to becoming 'B-E' as described earlier in this section – both squadrons utilising the same unit code simultaneously of course. *Author's collection*

Above: An anonymous MR.1 or 1A. *Author's collection*

Below: T.4 nose detail. *Author's collection*

MR.1A WB835 was involved in dropping trials of the Saunders-Roe Mk.3 lifeboat in 1951 and 1952. The boat, designed to be dropped from 700ft using four parachutes, measured 31ft by 7ft (at its widest point) and could accommodate up to ten people with provisions for two weeks. However, the airborne lifeboat concept soon gave way to the use of containerised Lindholme Gear which was accommodated inside the bomb bay. WB835 was scrapped in 1963. *Author's collection*

Shackleton MR.2

As good as the early Shackletons were, there was little doubt that improvements needed to be made. Therefore, in early 1951, VW126 was taken in hand to receive and test certain features that would appear on the forthcoming MR.2 – permission having been gained in July 1950 to begin exploring the next Shackleton variant. VW126's rebuild was completed by 19 July 1951, the date it first flew in its new configuration with an extended nose and repositioned radome amongst other changes.

The true MR.2 prototype was WB833, taken from the MR.1A production line, which first flew on 17 June 1952. In comparison to the MR.1, three external differences in particular were prominent: a much-extended nose accommodating twin 20mm Hispano cannon; a repositioned radome; and an extended tail cone. The extended nose featured a look-out position on its upper front section from where the two nose guns were operated remotely, while below it, a prone bomb aimer's position was provided with a (nearly) V-shaped, optically flat, front windscreen. The ASV 13 radar was moved much further aft to permit (with radome extended) a full 360° sweep that also made the equip-

ment itself less vulnerable to bird strike than was the case with a chin-mounted unit. The extended fuselage aft of the tail now included a transparent tail cone to provide a rear observation post for a prone crewman. Other, less obvious, external changes included the provision of a retractable tail wheel unit with twin wheels, and small ventral camera housings under the extreme rear fuselage.

Upgrades were ongoing throughout the MR.2's life, particularly the Phase Modifications, which primarily consisted of internal modifications that were not visible externally and thus, by necessity, largely fall outside the scope of this book, for example Phase I upgrades included the replacement of ASV 13 with ASV 21 radar. Phase II, however, introduced sealed exhaust units, plus new countermeasures which included an Orange Harvest S- and X-band radar-warning receiver (a distinctive mast, sometimes referred to as being 'sparkplug-like') which of course were visible externally. As for MR.2 Phase III airframes, the discernible external changes included the fitting of a new four-aperture, line-abreast flare dispenser on the starboard side of the fuselage, and cabin heaters adjacent to the radome.

Shackleton MR.2, WB833, was set aside from the MR.1 production line to become the prototype MR.2 which first flew on 17 June 1952. Several years of trials followed with various establishments before it was assigned to the ASWDU (coded 'B') in November 1960 for service trials. Following further modifications WB833 was allocated to 210 Squadron at the end of 1966 and received the code 'T'. Tragically, WB833's time with 210 Squadron was abruptly terminated on 19 April 1968 when, during an anti-submarine exercise, the aircraft struck the ground on the Mull of Kintyre in poor weather killing all eleven crewmen aboard. Author's collection

The first MR.2s were allocated for test and development purposes initially, consequently the first to be issued to the squadrons were WG555 and WG556 which went to 42 Squadron on 12 January 1953 with WG554 arriving on the 30th, just beating 206 Squadron which received WG558 on 3 February and WG557 on the 27th. Both these squadrons, in common with several

MR.2, WG531, illustrates the obvious external differences between an MR.1 and MR.2. First flown on 21 August 1952, WG531 went to 42 Squadron in April 1954 coded 'A-H'. Sadly tragedy struck on 11 January 1955 when WG531, Shackleton WL743 and all eighteen crewmen went missing south of Fastnet Rock, the presumption being that they had collided – two of five aircraft lost to the RAF that day, the others being a Meteor T.7, Spitfire 19 and a Harvard. *Author's collection*

Phase II MR.2, WG555 'N' of 204 Squadron is seen in September 1964 at Finningley. WG555 joined 204 Squadron in June 1961 with which it remained, except for modifications and a Phase III update, into 1971 when it was transferred to the Majunga Detachment Support Unit (MDSU) that formed on 1 March 1971 at Honington to supply MR.2s for the Beira patrol. (Reportedly five other Shackletons were also allocated: WL748, WL751, WL755, WL800 and WR961). A month later, on 1 April, 204 Squadron disbanded at Ballykelly, but on that day the MDSU was redesignated to become 204 Squadron at Honington. As the last maritime Shackleton squadron its duties, other than flying shipping-reconnaissance duties from Suffolk, included sending detachments to the Far East to take over 205 Squadron's responsibilities (utilizing some ex-205 aircraft) after it disbanded in October 1971. Recoded 'K' in 1967, WG555's final flight was to the RAF Fire Fighting School at Catterick in May 1972. *Author's collection*

others due to receive the new model, were also operating the earlier MR.1/1A as well – which soon led to problems.

The problem was that although the three Marks now in service were all Shackletons and all closely related, sufficient differences existed between them to make each different from the others in certain respects with operational and logistical problems arising particularly when MR.2s were operated alongside the earlier versions. (Actually this problem had been encountered before when the Lancaster B.4 and B.5 had to be renamed Lincoln B.1 and B.2 respectively to avoid logistical chaos!) Consequently it became necessary to implement a level of standardisation within the squadrons whereby each operated just one of the two basic Marks thus, during the middle of 1954, several units were required to relinquish their MR.2s after only a relatively short time – which helps explain several of the seemingly abundant 'irregularities' observed in Appendix 1.

The MR.2, as with the earlier models, was a very good maritime patrol aircraft, one that remained in service until the last MR.2 unit, 204 Squadron, disbanded in 1972. However, many of the issues concerning crew comfort in the MR.1/1A remained throughout with the MR.2 as well, one of the more uncomfortable being high levels of resonance created by the contra-rotating propellers that left crewmen fatigued and with a continual ringing in their ears. Thus production of the MR.2 was halted with WR969, the balance of the order being shifted to a redesigned Shackleton, the MR.3, one it was hoped would tackle and solve the issues of crew comfort as well as addressing other areas in which improvements were called for.

MR.2, WG557 'T-L' from 220 Squadron seen adjacent to an Avro Lincoln. WG557 had arrived with 220 on 19 March 1954, but was transferred to 228 Squadron six months later. *Author's collection*

WG557, by now 'L-L' with 228 Squadron seen in flight over Baginton, Coventry, in August 1955. *Newark Air Museum*

Following squadron service, WG557 received a series of modifications from November 1955 onwards followed by periods of storage. In December 1957, this airframe was allocated to the RAE at Farnborough for use in the development of fuzing systems for nuclear weapons as well as other duties. *Author's collection*

By now operated by the Empire Test Pilots School (ETPS), WG557 is seen at Farnborough on 12 September 1964 displaying its distinctive one-off colour scheme complete with red serial number and black 'Royal Navy' legend. By late 1964, WG557 was to be found on Farnborough's dump. *Courtesy of R.A. Scholefield*

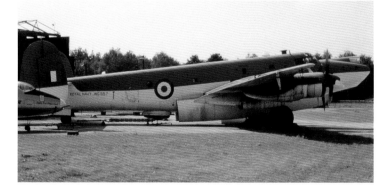

Transferred from 206 to 224 Squadron in August 1954, MR.2 WG558 was originally coded 'B-R' before it was modified to 'R', as seen here, dating the photo to *circa* October 1956. WG558 left 224 Squadron in July 1958 for updating followed by service with other squadrons and establishments. The white shield on the nose contains the Squadron's motif – a tower on a rock entwined by a serpent; its motto (translated) reads 'Faithful to a friend', referring to the Squadron's aid to Italy in WWI. *Author's collection*

Seen at Luqa, WG558 'Y' – by now a Phase II aircraft – had joined 210 Squadron in April 1963 and remained with the unit until November 1966. The motif on the fin is that of 210 Squadron which portrays a griffin, representing the unit's long association with Wales. *Via Roger Lindsay*

Seen with an apparent addition of a dorsal fin (it belongs to a neighbouring Neptune), Shackleton MR.2, WL737 'T-K' was photographed between March 1953 and July 1954 during its time with 220 Squadron. Thereafter, WL737 underwent several periods of modifications and repairs, as well as service with other units until October 1971 when it was deemed non-effective stock and was scrapped at St Athan in 1975.
Author's collection

WL740 arrived at Langar in March 1967 prior to being converted to a T.2. However, the work was halted and it was SOC early the following year and scrapped. The white disc on the nose is the Squadron motif incorporating a heron in flight, a bird that rarely misses its mark and one that will fight if attacked and relates to the unit's earlier history as a bomber squadron. *Author's collection*

Well-worn MR.2, WL741 'F', showing signs of having been to New Zealand or at least in contact with RNZAF personnel. In addition, albeit hard to appreciate, WL741 has white-painted upper wing surfaces extending to the wing tip. Having left 205 Squadron for Phase III modifications in 1966/67, WL741 returned to 205 in November 1967 when its code changed from 'H' to 'F', as seen here. This aircraft returned to the UK in February 1971 for storage prior to conversion and later delivery to 8 Squadron as an AEW.2. *Author's collection*

Close-up of WL741's tail section displaying 'zaps' of New Zealand and Australian origin. The Squadron motif, consisting of crossed trident and kris (sword), represent the unit's naval origins as well as its association with Malaya.

A poor quality image of an MR.1 and an MR.2 in flight, and while the distant MR.1 remains anonymous the other is MR.2 WL750. Presuming the letter 'T' represents a squadron code and not the aircraft then it's likely that this image dates from between July 1954 and June 1955, the only period in which WL750 was operated by 204 Squadron. Though 220 Squadron also used the code 'T', WL750 was never issued to that unit. *Author's collection*

MR.2, WL751 'B-L', 224 Squadron. This aircraft served with the unit from May 1953 until January 1959 when WL751 was dispatched to the manufacturer for Phase I modifications. Following further repairs, modifications, and service with several units including the MDSU/204 Squadron (see WG555), WL751 was withdrawn from service in 1972. *Author's collection*

Having been withdrawn from service it was hoped that WL751 might be preserved permanently once it had been acquired by Baginton-based 'Shackleton Aviation', where WL751 arrived in 1972. However, attempts to save it proved abortive and the aircraft was scrapped elsewhere in 1975. The part-word 'ETON', seen on the Company's offices in the background, forms part of the word 'Shackleton'. Note that in this instance the white upper wing stripe does not extend to the wing tip. *Fred Martin*

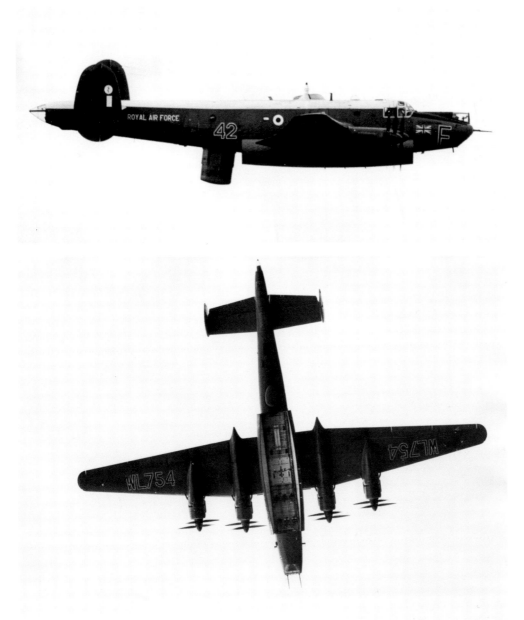

MR.2, WL754 'F', 42 Squadron, seen in September 1965. This aircraft was delivered to 42 Squadron on 16 September 1960, immediately after Phase I modifications had been completed. Phase II updates followed in 1962 with Phase III being undertaken from early 1966 to April 1967. On completion, WL754 went to 205 Squadron and was coded 'J'. Following service with other units, WL754 was placed in storage in April 1971 for eleven months following which it went to Bitteswell for conversion to an AEW.2.
Author's collection

Two views of WL754, seen in September 1965, is fitted with a distinctive 'sparkplug-like' mast on top of the fuselage which, familiar jokes to one side, is an Orange Harvest mast which was fitted to (some) Shackletons. This device was an 'S' and 'X' band radar warning receiver used to detect, classify, and ascertain the bearing of radar emissions operating within these specific bands.
Author's collection

A 'Suez era' Shackleton. MR.2, WL756 'T', of 38 Squadron seen wearing Suez Invasion stripes in late 1956. WL756 was later converted to an AEW.2. *Newark Air Museum*

Ballykelly Wing MR.2 'W' in November 1967. Centralised servicing by Coastal Command at Kinloss, St Mawgan, Ballykelly and Gibraltar, introduced from late 1965 to mid-1966, was primarily driven by the increased sharing of aircraft by squadrons when their own were away on detachment, although the removal of unit identities introduced a perception of squadron anonymity. (Exceptions at Ballykelly were the MR.3s of 203 Sqn and ASWDU Shackletons: they were flown only by their own crews.) *Author's collection*

MR.2, WL785 'E' of 37 Squadron wearing Suez Invasion stripes. *Roger Lindsay*

WL785 seen a few years later with 42 Squadron. The date of this image is unknown, but it cannot be earlier than July 1961, the month in which it returned to 42 Squadron following a Phase II upgrade. Previously coded 'C', it was only upon its post-modernisation return that WL785 received the code 'A'. WL785 remained with the Squadron until March 1965 – thus reducing the date envelope to a four year period. The motif on the fin includes the figure of Perseus and relates to the fact that 42 Squadron had been the first to use the Bristol Perseus engine. *Newark Air Museum*

WL786 was issued to 38 Squadron in mid-1959 shortly after receiving a Phase I upgrade and remained with the unit until mid-1961 when it was ferried to the manufacturers for further modifications. On completion, WL786 was issued to 205 Squadron in 1962 and remained with the unit until 4 November 1967 when harsh misfortune struck over the Indian Ocean and eight of the eleven men aboard were killed. A tragic sequence began with a mechanical failure and fire in the starboard outer engine which in turn prevented the propellers from being feathered. The engine fell away leaving the wing on fire, which led to part of the wing also breaking off – despite which the pilots were still able to ditch the machine. Unfortunately a heavy impact caused the fuselage to break apart and although a few crewmen were able to extricate themselves only three survived. *Author's collection*

First flown on 10 June 1953, WL789 was fitted with a magnetic anomaly detector in a tail boom (MAD boom) during July and August and arrived with ASWDU in September 1953 for MAD trials coded 'F-D'. It's unclear as to precisely how long trials continued, but it was at least four years, yet ultimately, despite attempts to resolve the matter, the magnetic anomaly detector proved electronically incompatible with other Shackleton electrical systems. The boom had been removed by April 1958 and WL789 was sent to an MU for upgrade and subsequent issue to the squadrons. *Roger Lindsay*

Following its MAD days, WL789 served with a number of units, including 224 Squadron. On completion of Phase II modifications in early February 1962, WL789 returned to the Squadron just days later (recoded 'A') and remained with the unit into October 1966. Following service with 38 and 205 Squadrons, WL789 was broken up in 1969. The disc on the nose contains 224 Squadron's rock and serpent motif. *Newark Air Museum*

WL796's first flight occurred in late August 1953 and within a couple of weeks was flying at Farnborough demonstrating the Mk.3 Lifeboat which was never actually dropped from an MR.2 nor, as related earlier, was it used by RAF Shackletons due to the introduction of Lindholme Gear. WL796 displays a less than distinct 'Hawker Siddeley Group' circular logo on its nose.
Author's collection

With three pairs of props feathered, WL796 demonstrates its prowess at Farnborough.
Author's collection

Following completion of Phase II modifications, WL796 was issued to 204 Squadron in May 1961 (coded 'M') with which it remained, despite any number of accidents, until transferred to 205 Squadron in October 1966. Seen between periods of accident repair, WL796 sits at Waddington in September 1964 displaying a full Squadron badge (not a motif) incorporating a cormorant perched on a mooring buoy – as inspired by a photo taken by a certain Aircraftman Shaw (aka Lawrence of Arabia). *Author's collection*

Waddington, September 1964. Transferred to 205 Squadron in October 1966, WL796 served in the Far East for a year before returning to the UK and disposal at Shawbury from November 1967. *Newark Air Museum*

Curiously, a 'wayward pixel' on the negative managed to alter WL797's serial into a spurious 'WL737'; fortunately the underwing serial is large enough to be unaffected by a single errant speck! Seen at Khormaksar, Aden, WL797 'C', had been allocated to 37 Squadron in April 1964 shortly after its Phase II modernisation. 37 Squadron disbanded on 7 September 1967, consequently WL797 was flown back to the UK a few days later and was sold for scrap the following year. *Newark Air Museum*

One of the great things about being a child living not too far from Langar was the lack of anything like a complete hedge, ditch or fence 'securing' the airfield; you could simply step through from an adjacent road with your dad to get a better view, and if your dad said it was okay to do so – it was! The airfield's security is rather different today of course. Having spent most of its life from December 1953 as a 38 Squadron aircraft coded 'X' throughout, MR.2 WL798 'X' is seen at Langar between May 1960 and April 1961 where it underwent a Phase II update. With its wingtip and starboard outer propellers missing, WL798 gives every impression that both the work and the aircraft had been abandoned, but such was not the case. Thus did 'X' became 'Y' with 38 Squadron in May 1961. In 1967 it returned to Langar for Phase III updates before joining 205 Squadron and later the Ballykelly Wing where its code advanced still further, becoming 'Z' in August 1970. By the end of the year WL798 had become 8114M at Cosford and later a source of spares for 8 Squadron's AEW.2s. *Author's collection*

Seen at Singapore, MR.2 WR952 was allocated to 205 Squadron at the end of January 1968, shortly after having received a Phase III update. Upon arrival it received the code 'B' which was retained, along with its 205 Squadron markings when this aircraft was transferred to 204 Squadron in late 1971. Returned to the UK in early 1972, WR952 was scrapped in 1973. *Author's collection*

Having joined 228 Squadron coded 'L-O' in 1954, it is thought that this image of MR.2, WR959, was taken between 1956 and 1958 when it was recoded 'W' and by which time the unit had ceased using their old code 'L' in favour of the Squadron number. Points of interest include the fuselage roundel outlined in white, while a lighter shade of paint has been applied to the upper surface of the nose. *Author's collection*

MR.2, WR959 'A' was transferred to 205 Squadron in mid-1966 and remained with the unit until it was broken up in Singapore from September 1968. *Author's collection*

MR.2, WR960 'X', photographed between August 1958, when the aircraft received its code, and March 1959 when it transferred to 42 Squadron and became 'B'. WR960 served with a number of units in later life until, in May 1971, it was sent to Bitteswell for conversion to the AEW role.
Author's collection

MR.2, Phase II, WR961 'S' was issued to 38 Squadron in May 1961 and remained with the unit until September 1966 when it departed for Langar and Phase III modifications.
Author's collection

WR961 went on to serve as part of the Majunga Detachment Support Unit in 1971, following which it returned to the UK and was stored at 5MU Kemble from 1972 and remained there until sold for scrap in 1978. Although this image isn't dated, it seems likely that it was taken at Kemble. *Fred Martin*

MR.2, WR962 of 228 Squadron, photographed between June 1958 (when it received the code 'Z') and June 1959 when it was issued to 204 Squadron and became 'R'.
Newark Air Museum

Having received a Phase III upgrade, WR962 was operated by 37 Squadron from March 1962 until the Squadron disbanded in September 1967.
Newark Air Museum

Seen at St Mawgan, WR962 was on its way to Shawbury on 9 September 1967. It was scrapped there in 1969.
Author's collection

Top: MR.2, WR963, arrived at Gibraltar to join 224 Squadron in October 1954 where it acquired the code 'B-M' and the unit's serpent-entwined rock motif on the fin. During 1956, WR963 was recoded 'M'. A survivor, this aircraft later went on to become an AEW.2 with 8 Squadron at Lossiemouth. *Author's collection*

Above: Following completion of Phase II modifications during 1961, MR.2 WR964 was allocated to 204 Squadron and received the code 'Q'. This airframe remained with the unit until February 1967 when it arrived at Langar, still coded 'Q', for conversion to T.2 standard and where it was photographed (from the road and outside of the boundary fence) still wearing 204 Squadron's motif – plus extensive exhaust staining. Conversion completed, WR964 arrived with MOTU in May 1968 and became 'Z'. At St Athan by mid-1970, it was struck off charge in late 1971 and scrapped at St Athan in 1978. *Author's collection*

Below: Phase II MR.2, WR966 'O', from 204 Squadron, photographed between June 1961 and December 1966 – the year it was flown to Langar for conversion to T.2 standard. In this image it can be seen that WR966 wears the Squadron's badge *and* motif (on the tail). *Newark Air Museum*

WR966 reverted to an MR.2 once its training equipment had been removed during November and December 1970, prior to being allocated to 205 Squadron as a Phase III MR.2 in January 1971 coded 'G'. In November 1971, WR966 joined 204 Squadron but retained 205's markings and returned to Britain in January 1972. This image, believed to have been taken in 1971 and in the UK (judging by the Hillman and Austin 1100 Countryman [estate] in the background), poses questions. Did WR966 embark on a return journey from Singapore to the UK and back during 1971, or does this image show the aircraft just prior to flying out to join 205 Squadron with that unit's markings and individual code already applied? Certainly, for its final journey back to the UK, in 1972, it was famously daubed with 'White Knuckle Airlines' and the semi-spurious code 'GT'. *Newark Air Museum*

MR.2, WR967 'F', from 42 Squadron photographed between March 1957 (when it joined the unit) and October 1960 when it was flown to Langar for Phase II modifications. Note the light coloured band around the fuselage (the significance of which remains unknown to the authors) and the exhaust stain on the nose. *Author's collection*

Following Phase II modifications, WR967 went to 38 Squadron as 'Z' in July 1961, but was recoded 'T' in June 1966. Returned to Langar in March 1967 for conversion to a T.2, on completion WR967 was issued to the Ballykelly Wing in September 1968 and later to 210 Squadron at Sharjah. With its useful life apparently at an end, it seemed, this aircraft was allocated to Catterick but instead was flown to Kinloss in November 1971 prior to 8 Squadron being reformed on 1 January 1972. Initially used as a crew trainer, WR967 was damaged in an accident nine months later following which, with wings now clipped, literally, the fuselage had been converted into an AEW.2 training aid by August 1975 to become the 'Dodo'. *Author's collection*

Left: MR.2 rear fuselage camera ports opened. *Author's collection*

Below left: MR.2 – the other end! *Author's collection*

Below right: MR.2 – 'twin twenties'. *Author's collection*

Bottom: An unidentified MR.2 from 42 Squadron drops a container to the mercantile *Weather Reporter*. Designed for the tough conditions of the North Atlantic, the ship had originally been the Castle class corvette HMS *Oakham Castle* until 1957 when she was sold to become a weather ship. She was scrapped in 1977 at Middlesbrough. *Author's collection*

Above: WG533 'D' 'White Knuckle Airlines' late of 204 Squadron (but retaining 205 Squadron markings on the fuselage). This image was taken at St Athan in July 1973 following its arrival back in the UK in early 1972. Assigned to the scrap man by the end of the year, WG533 was later dumped at St Athan instead. *Derek Hemingway via Fred Martin*

Below: WR966 'G' White Knuckle Airlines'. Struck off charge in June, WR966 was photographed at St Athan in August 1973 awaiting its appointment with the scrap man. The additional 'code letter' (T) was a zap while the figure '146800cc' applied to the combined cubic capacity of its four engines. *Derek Hemingway via Fred Martin*

Avros accumulate. Ansons, Shackletons (and a solitary Hastings) await their fate at Shawbury circa 1968. *Derek Hemingway via Fred Martin*

Shackleton AEW.2

rior to considering the MR.3, however, mention must first be made of a further Mark.2 variant – the Shackleton AEW.2, twelve of which ultimately entered RAF service at what would otherwise have been the very twilight of the Shackleton's career.

Although the Royal Navy had operated AEW aircraft for many years, it had become apparent by the late 1960s that the RAF too was also in need of a similar capability, albeit, to provide cover for the Navy's remaining ships as its aircraft carriers were progressively scrapped, along with its Gannet AEW.3s. For this new role and function, the most suitable and cost-effective airframe available was considered to be the Shackleton MR.2 – a type destined to be removed from service in the near future – providing that sufficient remained with reasonably low flying hours. A check was made across the fleet and twelve Phase III-modified airframes were selected, namely: WL741, WL745, WL747, WL754, WL756, WL757, WL790, WL793, WL795, WR960, WR963 and WR965.

Having been selected, the airframes were overhauled and then stored pending future conversion. The first to undergo the transformation from MR.2 to AEW.2 was WL745, which effectively became the prototype for the new Mark and which first flew in its new guise on 30 September 1971. Obviously, several internal changes were required for the new role, but externally the single most obvious physical change concerned the removal of the aft-mounted ASV radome and the incorporation of a forward-mounted radome encasing AN/APS-20F air-search radar.

It was decided that an ex-fighter unit, 8 Squadron, would be the one to operate the AEW.2. Having disbanded in late-1967, the unit reformed at Kinloss on 1 January 1972 in preparation for its new role, although it would be April before the first AEW.2, WL747, was received. The last to arrive was WL745, in September 1973, once its period of trials had been completed. Prior to the arrival of any AEW.2s, records indicate that two standard MR.2s, WL787 and WR967, were issued to the Squadron on 1 January 1972 as crew trainers – in fact at least three more MR.2 crew trainers would be used by the unit during the remainder of the decade, though not concurrently.

The Shackleton AEW.2 was only ever intended as an interim platform pending arrival of the Nimrod AEW.3, the development of which became ever more protracted with its in-service date being pushed further and further back, year-on-year, until, in

WL741 '41' seen at Finningley in September 1980. '41' joined 8 Squadron, the only unit to operate AEW Shackletons, in April 1973. Withdrawn from use in 1981, WL741 was flown to Manston where its remains were finally scrapped in 1987. *Fred Martin*

WL745 'O' was used to undertake a series of trials to test the type's compatibility with the venerable, but ageing, AN/APS-20 airborne early warning radar system, the first versions of which had equipped US Navy Avengers in March 1945. Shackletons, however, were equipped with the developed APS-20F system. WL745 effectively became the prototype AEW Shackleton and flew for the first time in its new guise on 30 September 1971. Both black and white images were taken a few days later and illustrate WL745's new AEW radome, blanked-off radar housing and temporary nose probe. *Both Author's collection*

1987, the decision was finally taken to abandon the programme entirely. In the meantime 8 Squadron's role of supplying AEW coverage to the RNv had expanded considerably to include radar coverage of UK airspace, directing air defence aircraft, electronic surveillance, plus other duties including the carriage of containerised Lindholme Gear in the Shackleton's bomb bay.

In 1981, following Secretary of State for Defence, John Nott's now infamous defence review, the number of AEW.2s was subsequently halved to leave WL747, WL756, WL757, WL790, WR963 and WR965, (initially 8 Squadron personnel only became aware of the decision because one of them happened to be watching the BBC evening news!) The six remaining aircraft would now have to suffice until the Nimrod AEW.3 entered service – which, as related, didn't happen and so they soldiered on until 30 June 1991 when they were replaced by the Boeing E-3D Sentry.

Following completion of the trial phase, and an upgrade to production standards, WL745 was issued to 8 Squadron on 17 September 1973. The last of twelve AEW.2s to join the unit, WL745 was withdrawn from use and flown to Catterick in July 1981. *Alan Carlaw*

First flown as an AEW.2 in January 1972, WL747 '47', arrived with 8 Squadron during April and remained with the unit until the end of the type's service with the RAF, making its last flight in July 1991 when it was flown to Cyprus for 'preservation'. *Alan Carlaw*

WL747 seen at Finningley in September 1976. 'Florence' (from 'Magic Roundabout' fame) is evident on the left-hand side of the nose. *Fred Martin*

This page: Three images of WL747 in May 1988. *Fred Martin*

Opposite page: Featured quite extensively in the MR.2 section, WL754 '54', now an AEW.2, was delivered to 8 Squadron in November 1972. Seen here in April 1980, WL754 was withdrawn the following year and flown to Valley for ground instructional use until finally scrapped in July 1987. *Fred Martin*

First flown as an AEW.2 in March 1972, WL756 was delivered to 8 Squadron two months later: it is seen here in September 1980. *Fred Martin*

Seen here in 1990 over-flying Leeming, WL756 was reportedly SOC in February 1988 according to three published serial listings and delivered to St Mawgan for ground instruction purposes shortly afterwards. In fact, WL756 remained with 8 Squadron into 1991, the year in which the last five Shackletons finally left RAF service. Ultimately WL756 did end up at St Mawgan where it lingered into 1999. *Fred Martin*

Seen at Coningsby in June 1975, WL757 '57' had joined 8 Squadron in late August 1972 and remained with the unit until SOC in 1991 at the end of the type's RAF career. WL757 was subsequently flown to Cyprus where, presumably, it continues to rot away with WL747. *Fred Martin*

WL757 seen at Lossiemouth circa mid-1970s. *Alan Carlaw*

WL757 as seen in October 1990. *Fred Martin*

Four images showing WL790 *circa* April 1990. WL790 joined 8 Squadron in September 1972 and, as many readers will be aware, this Shackleton was ultimately sold into preservation and now resides at Pima Air Museum, Tucson, Arizona, USA, registered appropriately as N790WL.
Top: *Fred Martin*; Centre: *Alan Carlaw;* Bottom left and right: *Newark Air Museum*

Two images of WL793 '93', seen in 1978. Withdrawn from use three years later, WL793 was on Lossiemouth's dump by mid-1981. *Author's collection; colour image Fred Martin*

Above and right: Seen *circa* July 1977, WL795 '95' had joined 8 Squadron in late 1972 and remained with the unit until withdrawn in 1981. It was flown to St Mawgan in November of that year where WL795 seemed destined to be used for firefighting practice, but that didn't happen and instead became a gate guard. As of December 2014 it was still in situ although it had by then been placed on the sale list by the MOD. Although sources state that this machine was named 'Rosalie', it is 'Zebedee' that can be seen in the black and white image. *Fred Martin & Alan Carlaw respectively*

Above and left: Seen in September 1979 and May 1982 (at Mildenhall) respectively, WR960 was taken on charge by 8 Squadron in June 1972 and served until withdrawn from service in November 1982. WR960 now resides at the Museum of Science and Industry (Manchester).
Both Fred Martin

Seen in July 1979, WR963 '63' was sold on 3 July 1991 and delivered to Coventry Airport six days later. *Fred Martin*

WR963 as seen *circa* 1989/1990. *Newark Air Museum*

AEW.2, WR965 '65'. Sadly, WR965 was lost on 30 April 1990 when the aircraft struck high ground, at about 800 feet, in thick cloud on the Isle of Harris killing all ten crewmen the last RAF Shackleton to be lost. *Author's collection*

The photographer's caption simply reads 'Eight of Eight at Lossiemouth'. Which is succinct enough.
Alan Carlaw

On short finals.
Alan Carlaw

This image was taken on 8 August 1988, or, as it states on the back of the photo – 'Squadron 8 on 8.8.88'. The aircraft involved were WL747, WL756, WL757 and WR963.
Author's collection

MR.2s CONTINUED TO SERVE TOO...

Right: In addition to the dozen AEW.2 conversions – and excluding non-flying MR.2s acquired for spares recovery – a number of MR.2s were allocated to 8 Squadron for pilot and crew training. It appears that at least five examples were assigned to the unit, albeit not all at the same time. Known examples were: WG556, WL738, WL787, WL801 and WR967 (latterly 'the Dodo'). Coded 'A' (because there was already a '56'), WG556 is seen in March 1978. Requiring an overhaul after six years of open storage at St Athan, this machine was delivered to 8 Squadron in May 1977 to replace WL738 as its fatigue life had almost expired. WG556 lasted into 1980, the year in which it suffered a heavy landing and was thus relegated to ground instructional duties before going to the fire section. It was scrapped in 1982.

Above: WL738, '38', seen at the Lossiemouth Air Day on 13 August 1977, with WL795 beyond. WL738 later became a gate guardian at Lossiemouth after having served 8 Squadron from March 1974 until October 1977 when its fatigue life expired. Placed on the gate six months later, WL738 was scrapped in the 1990s. *Both Alan Carlaw*

Above: MR.2 crew trainer, WL801 '01', had been allocated to 8 Squadron in August 1974 and served until its fatigue life was all but expired. Seen here in May 1982, WL801 had been flown to Cosford in June 1979 for use as an instructional airframe. Although later selected for preservation, it wasn't to be and WL801 was scrapped at Cosford *circa* 1985. *Fred Martin*

Left: 'Brian' (the snail) was a 'Magic Roundabout' character. His was one of several 'personalities' to appear in a popular TV show in the 1970s whose individual names were applied to 8 Squadron's Shackletons for a while. Some airframes were later renamed – no doubt a 'devious ploy' intended to sow confusion among us decades later. 'Brian' (also known as WL757) was photographed at Leuchars on 15 September 1973. *Alan Carlaw*

Shackleton MR.3

The prototype MR.3, WR970, was first flown on 2 September 1955 powered, as were all MR.3s (and, retrospectively, MR.2 Phase III airframes too) by Griffon 58 engines. Amongst many changes the new Mark featured a revised nose, tricycle undercarriage (dictating a shortened bomb bay) with twin main and nose wheels, permanent tip tanks each containing approximately 253 gallons of fuel, noise-reducing sealed exhaust units and new cockpit glazing with significantly less framing than previous Marks for improved pilot visibility. This time the fuselage was fitted with sound-deadening materials (also applied retrospectively to MR.2 Phase II airframes), as well as more comfortable seating to help improve crew comfort.

No 220 Squadron (renumbered 201 Squadron on 1 October 1958) became the first operational unit to receive the MR.3 following the delivery of WR976 on 30 August 1957; thereafter MR.3s were delivered to 42, 120 203 and 206 Squadrons. Excluding foreign sales, total deliveries of the MR.3 amounted to thirty-four airframes including prototype WR970 which crashed while undertaking stall tests over Derbyshire on 7 December 1956.

During its service life, the MR.3 also underwent Phase upgrades that included the installation of Orange Harvest and other sensors. But one Phase III modification went much further, it being a result of the increased weight of MR.3s generally, and Phase III airframes particularly, which, even *before* they received auxiliary engines, weighed, at maximum take-off weight (about 108,000lbs), approximately 12,000lbs more than MR.2 Phase III airframes. Under certain conditions, particularly in hotter climates, the power of the four Griffon 58s was insufficient to ensure the aircraft left the ground safely within a given distance, a problem exacerbated by the fact that it was not possible to further boost the output of the Griffons. Hence the decision was made to strengthen the wing and install two Bristol Siddeley Viper 203 auxiliary turbo-jet engines, one in each outer engine nacelle to supplement the power of the four Griffons, with each jet providing about 2,700lbs of thrust. Surprisingly to some perhaps, the jets ran on AVGAS (petrol) which was okay providing they were only used for short durations at *full* power. With a dry weight of just 549lb, the Viper 203 developed a thrust-to-weight ratio of 4.9: 1, its exhaust being vented through an outlet at the rear of the modified outer nacelles, while the jets' air intakes hinged down underneath the

The first of its Mark, prototype Shackleton MR.3, WR970, was first flown on 2 September 1955 and is seen here a few days later at Farnborough.
Newark Air Museum

WR970 is seen with a nose-mounted probe and yaw meter. During trials conducted at Boscombe Down from September 1956, WR970's stall characteristics were found to be unacceptable and the prototype was later returned to the manufacturers to have improved stall-warning devices fitted. Sadly, on 7 December 1956, the aircraft crashed in Derbyshire while undertaking stall tests killing all four crewmen aboard. *Author's collection*

nacelles when the jets were running. The last MR.3 to receive Vipers was XF707, its jet installation being completed in January 1968.

The last UK-based MR.3s belonged to 42 Squadron which relinquished them in September 1971, while 203 Squadron, in Malta, retained theirs into December, most of which were returned to the UK in January 1972 for storage or disposal.

Although the Shackleton MR.1 and MR.2 never attracted any foreign sales, the MR.3 did. In March 1954 the South African government placed an order for MR.3s to be built and supplied to Phase I standard. The first pair, from a total of eight on order, were handed over to personnel from 35 Squadron South African Air Force (SAAF) in the UK in May 1957 and were flown out to South Africa three months later, with the remaining half-

dozen arriving periodically over the next several months. South Africa's Shackletons received the serials 1716 to 1723.

Due to an embargo imposed by the United Nations over South Africa's policy of apartheid, acquiring components for the Shackleton fleet became increasingly difficult and thus the aircraft's serviceability suffered. The fleet had been modified to Phase III standards prior to the implementation of the arms embargo, albeit without auxiliary Viper engines. Two of the aircraft were re-sparred, 1716 in the UK, and 1717 in South Africa, but the lack of engine spares and tyres, together with airframe fatigue, took its toll and by November 1984, the fatigue lives of all but the two re-sparred aircraft had expired and the fleet was retired into storage, with the exception of 1718 which had been destroyed in a crash in 1963.

With the loss of the prototype, WR971 (first flown in late-May 1956) was used for much of the type's development work and it wasn't until October 1960 that it was finally allocated to 120 Squadron, its first RAF unit. Later updated to a Phase III aircraft, it became 8119M *circa* 1971. *Author's collection*

Another view of WR971, this time seen over the 'Needles' wearing a different-coloured fuselage serial number. *Author's collection*

WR972 seen during its days with the RAE who obtained it in March 1959 – about two and a half years after its first flight. As can be seen in this post-April 1968 image, it was fitted with a rear-fuselage observation 'turret'. *Author's collection*

In March or April 1968, WR972 was painted in a style that complied with the RAE's standard colour scheme as seen in this September 1968 image taken at Coltishall. WR972 was SOC at Farnborough on 31 January 1973 and used for fire practice. *Fred Martin*

MR.3, WR975 'F', from 203 Squadron photographed on 17 September 1960. First flown in mid-1957, this aircraft had received a Phase I upgrade in 1959 and was issued to 203 Squadron in November that year. Phase II and III updates followed in 1961 and 1963 respectively. Following service with several other units, WR975 was scrapped at St Athan in 1971. *Roger Lindsay*

MR.3, WR976 'M', a 201 Squadron machine, is seen in September 1965 five months after receiving Phase III modifications. In 1966 it was fitted with auxiliary jet engines (Vipers) and in early 1967 was allocated to the Kinloss Wing. Tragedy struck during a maritime exercise in the vicinity of the Scilly Isles on 19 November 1967 when WR976 hit the sea and exploded killing nine of the eleven crew.
Author's collection

MR.3, WR979 'D' was first flown in November 1957, although this image was taken in 1970 virtually at the end of its operational life. When WR979's Phase III modifications were completed in September 1965, the aircraft joined 120 Squadron and received the code 'D' which was retained when WR979 became part of the Kinloss Wing – a unit pooling arrangement introduced from 1966. In late 1968, WR979 received Viper jet engines, as seen in this image, and remained with the Wing until the end of July 1970 when it was flown to St Athan and stored. It was SOC in October 1971 and scrapped soon afterwards.
Newark Air Museum

MR.3, WR981 'A'. Photographed in 1958/59, 206 Squadron was the first unit to which this aircraft was allocated following its first flight in December 1957. Thereafter, WR981 went on to serve with several units, received several repairs and several modifications until finally being flown to St Athan in November 1970 to be stored. However, within a month the aircraft was flown to Topcliffe where it became 8120M. In this image the Squadron motif is seen within a white disc on its fin depicting an octopus – the unit's motto being 'Nought escapes us'.
Newark Air Museum

WR981 (8120M) 'G', late of the Kinloss Wing and seen at Topcliffe in July 1975, still displays its original serial number. WR981 was scrapped at this location ten years later. *Fred Martin*

MR.3, WR989 'B', from 120 Squadron. First flown in June 1958, this machine was allocated to the unit two months later and remained with it, other than departures for upgrading, until December 1963. *Author's collection*

WR989 'B', still with 120 Squadron, albeit with small detail changes. *Author's collection*

WR989 was allotted to 201 Squadron following completion of Phase III modifications in mid-1965, the unit having allocated the code 'K' to this airframe at about that time. As can be seen, WR989 had yet to receive Viper jet engines which were subsequently fitted in the first few months of 1967. *Author's collection*

With Vipers installed, WR989, seen *circa* summer 1968, was serving with the Kinloss Wing while retaining the code 'K' from earlier times. Transferred to 203 Squadron in early 1970, WR989 remained with that unit until early 1972 when it was placed in storage prior to being despatched to Leeming for fire-fighting practice later that year, and where it was scrapped in 1975. *Author's collection*

MR.3, XF700 'M', a Phase III-modified aircraft seen at Leuchars in 1968 complete with Viper jet engines which had been fitted the year previously. At this time the aircraft was being operated by the Kinloss Wing and remained with the unit until transferred to 203 Squadron in early 1969 coded 'F'. XF700's last flight occurred in October 1971 when it was flown to Nicosia, Cyprus, for use by the fire section and where its carcass is still believed to exist today. *Roger Lindsay*

MR.3, XF703 was issued to 203 Squadron in January 1959, three months after its first flight, and served with the unit for a year before being despatched to receive Phase I modifications. This machine would later return to 203 Squadron in 1961, when it was coded 'L', thus the apparent absence of any code in this image suggests the photograph was taken in 1959. The faint silhouette on the aircraft's nose depicts the winged seahorse that appears on the Squadron badge. XF703 was allocated to the RAF Museum store at Henlow in September 1971 for preservation, but was scrapped there instead in 1975. *Newark Air Museum*

MR.3, XF704 'L', from 201 Squadron as seen in summer 1965 shortly after its Phase III modifications had been completed and prior to Viper jets being fitted. In fact the latter never would be as XF704 crashed into the Moray Firth, eight miles from Kinloss, on 8 December that year killing all seven on board: it is believed that the aircraft may have stalled during its approach to the airfield. *Author's collection*

Having received Phase III modifications, XF707 was issued to 42 Squadron in January 1966 receiving the code 'D'. This MR.3 would be the last of its type to receive Viper jets engines which were installed during the second half of 1967, thus dating this image to within an eighteen-month period as the Vipers had yet to be fitted when this photo was taken. A minute 42 Squadron badge can be seen on a white panel on the nose. XF707 remained with 42 Squadron until withdrawn from operational use in early 1971 and was later delivered to Topcliffe where it was scrapped in 1977. *Newark Air Museum*

Long since withdrawn from flying duties, XF707 'D' (repeated on tail cone) is seen at Topcliffe with the hulk of an Argosy lying beyond. *Fred Martin*

This image of MR.3, XF708 'O', late of 201 Squadron, was taken at Langar on 7 July 1962. It had arrived two months earlier to receive Phase II modifications which were completed in July 1963. Thereafter, this aircraft served with 120 and 203 Squadrons until January 1972. Today, as many will know, XF708 resides at Duxford where it arrived on 23 August 1972 – the date of its last flight. *Author's collection*

Phase I MR.3, XF709 'N', seen on an unspecified date between mid-1959 and July 1962, the only period in which it carried this particular code. XF709 was subsequently allotted to other units, the last being the Kinloss Wing, with which it remained until mid-1970 when it was flown to St Athan where it was scrapped in 1972. *Roger Lindsay*

MR.3, XF710 'K', from 201 Squadron seen between April 1959 and August 1962. Operated later by 120 Squadron, on 10 January 1964, XF710 was sent to track a Soviet submarine operating off of Northern Ireland. Shortly after taking off from Kinloss, problems developed with one of its starboard engines that in turn led to an uncontrollable fire developing in the wing necessitating a crash-landing on Culloden Moor near Inverness, which was successfully completed without fatalities. *Newark Air Museum*

MR.3 MISCELLANY

Opposite, top: A low flyby. *Author's collection*

Centre and bottom: Three views of an unidentified MR.3 with Vipers installed. *Author's collection*

Opposite page:
Although unidentified, there is a high probability that this MR.3 is WR978 which was issued to 220 in November 1957 and coded 'M'. 220 Squadron only used the type from August 1957 to October 1958 when it became 201 Squadron. *Author's collection*

An unidentified MR.3 from 206 Squadron. *Author's collection*

MR.3, 1717 'O'. One of the first three Shackletons delivered to the South African Air Force (SAAF) in the Dark Sea Grey and PRU Blue scheme with 'springbok' roundels. *Martin Blundell*

This page:
MR.3, 1717 'O', SAAF, aft fuselage close-up. *Martin Blundell*

MR.3, 1721 'N', SAAF, showing its twin 20mm cannon and the last two characters of the aircraft's serial number. *Martin Blundell*

MR.3, 1722 'P'. In service, SAAF Shackletons were operated by 35 Squadron which employed yellow individual aircraft letters in the range 'J' to 'Q'. *Martin Blundell*

Shackleton Camouflage and Markings

Royal Air Force

When the Avro Shackleton entered service with RAF Coastal Command in 1951 it did so finished in the then standard camouflage scheme for Coastal Aeroplanes as laid down in Air Publication (AP) 970 'Design Requirements for Service Aircraft' Chapter 108, 'Camouflage and Aeroplane Identification Markings' Amendment List (AL) 44 dated 1 February 1949. This stated that medium and long range anti-shipping, anti-submarine and maritime reconnaissance aeroplanes were to be finished on most of their upper surfaces in Medium Sea Grey. The under surfaces, to Pattern No.1, (i.e. with the demarcation high up on the fuselage sides) were to be gloss White. The under surfaces between the boundaries of Pattern No.1 and Pattern No.2 (i.e. demarcation low down on the fuselage sides) were to be matt White. In addition to this, the matt White was to be extended upwards and merged into the upper surfaces in such a manner that when viewed in side and front elevation the aeroplane was to appear almost entirely white. On the engine nacelles, the white finish was to be extended upwards to cover the whole of the top surface forward of the boundary of the upper surface colour of the wing near the leading edge.

Standard post-war 1-2-3 proportioned national markings in Bright Red, White and Bright Blue were to be applied to the upper surfaces of the wings and sides of the fuselage. The upper wing roundels were of 84 inch overall diameter whilst those on the fuselage were of 54 inch diameter. An equally proportioned fin flash, 36 inches long and 24 inches high, was to be applied to both sides of both fins.

The serial number was to be applied to the under surface of the wings in black characters 48 inches high and to the rear fuselage in Light Slate Grey characters 8 inches high. The underwing serial numbers were applied in the traditional manner, outboard of the engine nacelles, reading from the rear under the port wing and from the front under the starboard wing.

The Medium Sea Grey and White scheme for maritime aircraft as described above was retained following a routine camouflage policy review which concluded in October 1950 and was included in Air Ministry Order (AMO) A.217 dated 19 April 1951. This scheme was consequently applied to all Shackleton MR.1 and MR.2s on the production line.

Squadron markings

In March 1951, the RAF revised its long-standing squadron marking policy and decided to abandon the two-letter 'squadron code' identification system which had been in use since 1938. On 7 March 1951, Headquarters Coastal Command wrote to its Groups to inform them of this decision and that the changeover to a new system was to be completed by 30 June 1951.

In Coastal Command, each aircraft would now carry two letters. The first was a squadron identification letter which was to be allocated on a station basis to identify which squadron on a station the aircraft belonged to and was to be placed aft of the fuselage roundel. For operational units of all commands the choice was restricted to 'A', 'B', 'L' or 'T'. For operational training units (OTUs) and other non-operational units of Bomber, Coastal, Fighter and Transport Commands, the letters 'C' to 'K' inclusive (less 'E' and 'I') were to be used.

The second letter identified an individual aircraft within a squadron and was to be placed forward of the fuselage roundel. Both letters were to be applied in accordance the instructions for the application of unit code letters in AP 2656A 'Camouflage and Marking of Aircraft' Volume 1, Section 6, Chapter 2, which stated that such markings were to be Light Slate Grey on Coastal Command aircraft which had White under surfaces.

In addition to these code letters, RAF squadrons were also permitted to display their squadron badges on aircraft in accordance with the regulations laid down in AMO A.334/50 dated 18 May 1950. This stated that at the discretion of the Air Officer Commanding-in-Chief (AOC-in-C), badges could be displayed on aircraft either complete with the standard frame and motto, provided that the overall height did not exceed 18 inches, or without the frame or motto on a circular white background 18 inches in diameter with a half inch border of pale blue. AMO A.232 dated 10 September 1953 modified the design of the standard frame to show a Queen's Crown in place of a King's Crown following the Coronation of HM Queen Elizabeth II in June of that year.

From August 1953, HQ Coastal Command decreed that on Shackleton aircraft the individual aircraft code letter which was to be placed forward of the roundel would in future be marked on each side of the nose of the aircraft. On 10 August a drawing was circulated to Coastal Command units showing the position of these markings and stating that the letters were to be in accordance with AP 2656A. This was followed on 8 October 1953 by a revised drawing which showed a new layout which also allowed for a squadron badge to be marked on the nose just aft of the letter.

Revised Camouflage Finish

On 28 July 1954, HQ 18 Group wrote to HQ Coastal Command to state a case concerning the most practical colour scheme for the Shackletons of 236 OCU. It had been found that these hard-flown aircraft became extremely dirty and were both difficult and time consuming to clean. The resulting contrast in cleanliness between the Shackletons and other aircraft was often remarked on to the detriment of Coastal Command's reputation.

HQ Coastal Command wrote to the Air Ministry in a letter dated 2 September 1954 outlining the problem and informing the Air Ministry that the AOC-in-C had decided to exercise his prerogative under paragraph 27 of AMO A.658/52 and to authorise 236 OCU

Above: Shackleton MR.1, VP258 photographed while serving with 120 Squadron in the early/mid-1950s, in the medium and long range anti-shipping, anti-submarine and maritime reconnaissance aeroplane Medium Sea Grey and White scheme to Pattern No.1. The unit code 'A' and the individual aircraft letter 'C' were painted in Light Slate Grey, as was the fuselage serial number with underwing serials in black. The aircraft displays 120 Squadron's badge on its nose – an Icelandic falcon perched on a globe. *Author's collection*

Below: Shackleton MR.1A, WB834 from 236 OCU *circa* 1955 in the then newly introduced overall gloss Dark Sea Grey finish. Standard post-war national markings in Bright Red, White and Bright Blue were retained, but squadron and individual aircraft code letters and serial numbers were re-applied in white as specified in AP 2656A for aircraft wearing a dark camouflage scheme. Although the squadron code letter 'C' remained on the fuselage, the individual aircraft letter 'L' was applied to the nose. *Author's collection*

to modify the standard finish of their Shackletons from the standard white to a blue/grey colour.

Whilst the repainting of the Shackletons of 236 OCU was accepted by the Air Ministry in a letter dated 22 September 1954, a proposal put forward by HQ Coastal Command on 18 October 1954 that all Coastal Command's Shackletons should adopt a similar finish met with stubborn resistance and there followed a war of words between AOC-in-C Coastal Command and the Air Ministry which was finally resolved in Coastal Command's favour on 28 June 1955 when the Air Ministry wrote to HQ Coastal Command giving consent for the adoption of overall gloss Dark Sea Grey on the Shackleton fleet.

Following the application of the new gloss Dark Sea Grey finish, code letters and serial numbers were applied in White as specified in AP 2656A for aircraft wearing a dark camouflage scheme. Subsequently, the new glossy finish appears to have caused some problems for the pilots with light being reflected from the forward fuselage and this led to the adoption of a matt black anti-glare panel which extended over the upper surface of the fuselage forward of the windscreen to varying degrees. This feature became common to most Shackletons until the type passed from service.

Squadron Numbers

In January 1955, the Air Ministry lifted security restrictions concerning the use of squadron numbers with the object of identifying squadrons more closely with the localities near to their bases. Following this, in May 1955, HQ Coastal Command requested that consideration be given to permitting squadrons to paint the squadron number on the sides of their aircraft. It was thought that the adoption of such a practice would establish the identity of squadrons not only in their local community but also with the general public and also have the added benefit of raising the profile of a squadron within the RAF which would have a positive effect on its morale.

Evidently some kind of interim permission must have been given for the application of squadron numbers to aircraft as when four Shackleton MR.2s of 228 Squadron set out on Operation *Suntan*, a goodwill tour of South American countries in October 1955, they carried their Squadron number '228' on the rear fuselage in Bright Red numerals with a thin white outline. Following this tour, on 7 November 1955, HQ Coastal Command wrote to the Air Ministry to request early confirmation that squadron numbers could be painted on aircraft as standard markings. In their reply dated 14 December 1955, the Air Ministry granted permission for Coastal Command to apply squadron numbers to their aircraft.

On 11 January 1956, HQ Coastal Command wrote to RAF St Eval requesting details of how the squadron number had been applied to the Shackletons of 228 Squadron before notifying its subordinate units in a letter dated 29 March 1956, that the Air Ministry had given approval for squadron numbers to be painted on the aircraft of Coastal Command and that it had been decided that all letters and numbers were to be painted red, and, with the exception of fuselage serial numbers, outlined in white. Stations were to paint squadron numbers on their Shackletons and re-colour existing identification letters and serial numbers as soon as possible. At the same time, the rescue 'break-in' markings were to be altered from Bright Red to Yellow.

Suez markings

The distinctive markings applied to allied aircraft during Operation *Musketeer*, the invasion of the Suez Canal area of Egypt in November 1956 were described in Appendix 'L' to the overall Air Plan under the heading 'Terrapin'. This stated that all Allied aircraft taking part in the operation including those allotted for air defence, but excluding Vickers Valiants and helicopters, were to be marked with black and yellow stripes. The markings were to consist of three yellow and two black stripes which alternated around the centre of each wing and around the fuselage or tailboom just forward of the tailplane. For Canberras, transport and maritime aircraft such as the Shackleton, each black or yellow stripe was to be 24 inches wide. Shackletons of both 37 and 38 Squadrons, operating from Malta, adopted these markings with effect from D minus 1.

Colonial Policing Detachment and Operation *Grapple* 1956-1958

Just prior to 42 Squadron's detachment to Aden to assume a colonial policing role in early December 1956, HQ Coastal Command informed the Air Ministry that in view of the climate in Aden, and out of concern for the ground and air crew who would have to work in these conditions, it had been decided to paint the upper surface of the fuselage on the first four aircraft of the detachment in glossy White, so as to reflect some of the heat and to re-evaluate the situation for the replacement aircraft.

On the other side of the world at about the same time, Coastal Command was also making a contribution to Operation *Grapple*, the British thermo-nuclear bomb tests which were conducted at Christmas Island in the Pacific between 1956 and 1958. It would appear that during the initial 206 Squadron detachment to Christmas Island in 1956-57, the benefits of painting the upper surface of the fuselage white to reflect the heat also became apparent.

In the light of the experience gained in these two detachments, on 18 March 1957, HQ Coastal Command informed the Air Ministry that it had been decided to also paint the upper surface of the fuselages of successive 42 Squadron replacement Shackletons. Further to this, the Shackleton Detachment commander at Christmas Island had strongly recommended that the Shackletons engaged on Operation *Grapple* be similarly painted and HQ Coastal Command had therefore decided to paint the upper surface of the fuselages of all *Grapple* Shackletons white.

Besides the white upper surface to the fuselages, some *Grapple* Shackletons also carried the Operation *Grapple* badge which consisted of a red frigatebird carrying a grappling hook applied either to the fuselage or the outer faces of the fins.

When 37 Squadron moved permanently to Aden for the colonial policing role in August 1957, their Shackletons had white tops extending further down the fuselage sides than became standard on the rest of the Shackleton fleet from 1959 with the bottom edge coinciding with the bottom edge of the cockpit canopy.

For the 1958 deployment of 269 Squadron's Shackleton MR.1s for the *Grapple Z* trials, their basic Dark Sea Grey colour scheme was modified to incorporate a heat reflecting white finish on the radome and upper surface of the fuselage. The mainplanes had Bright Red wing tips from the aileron hinge line out to the tip on both the upper and under surfaces

Above: The overall glossy Dark Sea Grey scheme was retained by the Shackleton fleet throughout the rest of its service with the RAF. This MR.3, XF707, of 201 Squadron illustrates the next change of markings with the introduction, in 1956, of squadron numbers applied to the rear fuselage and an individual aircraft letter on the nose which replaced the earlier unit code letter system. At the same time it was decided that all letters and numbers were to be painted red outlined in white – except fuselage serial numbers which would be red only. The upper surface of the fuselage was also painted white, to reflect heat on aircraft operating in hot climes initially, but later to all Shackletons (excluding AEW.2s) in the interests of standardisation. *Newark Air Museum*

Below: Number 8 Squadron reformed at Kinloss as an AEW unit in January 1972 equipped with Shackleton AEW.2s such as WL790 photographed here *circa* April 1990. Finished in the overall glossy Dark Sea Grey scheme, WL790 features the Squadron's traditional colours of sand, blue and red applied in bars flanking the fuselage roundel. From the early 1960s the legend 'ROYAL AIR FORCE' was applied to the rear fuselage of Shackletons just forward of the tailplane in white upper case characters some 12 inches high, which led to squadron numbers (numerals) being reduced in size and moved forward. Similarly the fuselage roundel, was also reduced in size and moved forward to just above the trailing edge of the wing. The glossy Dark Sea Grey finish appears to have caused some problems for the pilots with light being reflected from the forward fuselage which led to the adoption of a matt black anti-glare panel forward of the windscreen. Underwing roundels were applied to AEW.2s and in place of an individual aircraft letter, the last two characters of the serial number were applied in red, outlined in white, above the flash on the outer faces of the fins. *Alan Carlaw*

whilst both surfaces of the tailplane were similarly treated. 269 Squadron also employed Bright Red spinners. The fin markings were modified to carry a Union Flag on the outboard faces of the fins in place of the usual fin flash which was retained on the inner faces. This was an established practice on Shackletons spending time at foreign airfields outside the UK, either on goodwill tours or extended operations in company with other NATO aircraft because experience had shown that the standard flash was often mistaken for the French national flag.

Grey Aircraft

Following the decision of 28 June 1955 that all Coastal Command Shackletons were to be finished in Dark Sea Grey, the Ministry of Supply was asked to arrange for the new Shackleton MR.3s then on order to be finished in this scheme on the production line. As a consequence, RAF Shackleton MR.3s were finished in overall Dark Sea Grey from the start of their service in 1957.

AMO A.24/58, dated 15 January 1958, saw a wholesale revision of the classification of RAF camouflage schemes so that they were listed by colour rather than operational role. Paragraph 7 was entitled 'Grey Aircraft' and stated that maritime reconnaissance aircraft and meteorological aircraft were to have a high-gloss finish in Dark Sea Grey. The serial number was to be red and the aircraft identification and unit markings red edged in white. This basic scheme could be modified to meet local conditions on the authority of AOC-in-C or C-in-C provided that the Air Ministry was informed and the changes were met from Command resources.

One such change was promulgated in AMO A.285/59 dated 9 December 1959 by which time the white finish on the upper surface of the fuselage had been extended to the whole of the Shackleton fleet in the interests of standardisation. AMO A.285/59 extended the heat reflecting white finish to also cover the area of the wing fuel tanks. This instruction was subsequently amended in AMO A.239/60 dated 23 November 1960 which stated that the white finish over the area of the wing fuel tanks was only applicable to Far East Air Force (FEAF) aircraft.

New Markings, Organisation and Nomenclature

In 1961 an incident occurred in the Red Sea in which a Soviet freighter was allegedly 'buzzed' in an aggressive manner by a maritime reconnaissance aircraft which was apparently incorrectly identified as an RAF Shackleton. Following this incident, the AOC-in-C Coastal Command expressed a wish that the legend 'ROYAL AIR FORCE' be applied to the side of all Shackleton aircraft. "… to avoid confusion with aircraft of similar types operated by other nations".

HQ Coastal Command informed its units of the new policy in a letter dated 4 September 1961. The application of the legend 'ROYAL AIR FORCE' to the rear fuselage just forward of the tailplane in white upper case characters some 12 inches high which extended for a distance of approximately 10 feet along the side of the fuselage led to the squadron number being reduced in size and being moved forward to a new position just aft of the mainplane above the ventral radome. The fuselage roundel was also reduced in size, down to 36 inches in diameter and moved forward to just above the trailing edge of the wing. The Bright Red serial number was placed centrally underneath the 'ROYAL AIR FORCE' legend.

1964 saw the amalgamation of the Admiralty, War Office and Air Ministry into the Ministry of Defence (MoD) and from this point the old practice of issuing aircraft camouflage and marking instructions to the RAF by AMO ceased and was replaced by the issue of Defence Council Instructions (RAF) hereafter referred to as DCIs.

Also during 1964, the aircraft camouflage and marking colours commonly used by the RAF were incorporated into British Standard 381C and from this date, APs and DCIs begin to define the colours being quoted by a BS 381C number in addition to the colour name. As far as the Shackleton was concerned, these common colours were Dark Sea Grey BS 381C/638, the basic overall camouflage colour; Post Office Red BS 381C/538 and Roundel Blue BS 381C/110, the colours used in the national markings which had previously been referred to as Bright Red and Bright Blue respectively, and Golden Yellow BS 381C/636 which was used for wing walkway markings, the propeller tip markings and an assortment of other warning notices. Neither Black nor White were ever included in BS381C so no such number is ever quoted for these colours.

DCI T.346/65 dated 4 August 1965 extended the practice of applying the heat-reflecting white finish over the area of the wing fuel tanks to aircraft of Air Force Middle East (AFME) in addition to those of FEAF. This instruction was incorporated in AP 970 by AL.97 in March 1966.

Centralised Servicing

From 1966, the advent of centralised servicing in the UK saw the deletion of squadron numbers as Shackletons were 'pooled' within a Wing organisation with only individual aircraft letters being applied. In 1968, the Kinloss Wing painted a repeat of the aircraft letter in Dayglo orange on the rear of the fuselage immediately forward of the tail cone whilst the St Mawgan Wing applied similar markings in white. At about the same time small station badges inside standard frames began to be applied to home-based 'pooled' Wing aircraft whilst, where only a single squadron was based at a station in overseas commands, squadron badges began to re-appear.

During the autumn of 1969, the tips of the propeller blades changed colour from the yellow tips which had been in use since 1937, to feature red-white-red bands. This is thought to have been to meet the requirements of a new NATO Standard which had its origins in the adoption of red-white-red propeller tip markings in Canada in 1968 on the grounds that such markings were more easily visible than the yellow tip-marking previously used. The new colour scheme for the propellers was included in DCI S.136/69 13 August 1969 which stated that the tips of propellers were to be coloured with bands of Post Office Red-White-Post Office Red. On the Shackleton, these markings were applied to both the front and rear faces of the propeller blades.

Shackleton AEW.2

No 8 Squadron reformed at Kinloss for the AEW role on 1 January 1972, equipped for crew training with two standard Shackleton MR.2s. Initially, the Squadron applied its squadron markings in the form of its traditional Squadron colours of sand, blue and red bars flanking the Squadron's Arabian dagger motif on a white circular field with a thin pale blue surround in the usual place on the nose immediately aft of the aircraft letter.

With the advent of the BAe Nimrod and the Shackleton AEW.2 a change was made to the nomenclature used in DCI S.70/72 which was promulgated on 26 April 1972. Paragraph 12 was headed 'Maritime Reconnaissance and Airborne Early Warning Aircraft' and was split into two sub-paragraphs. The first of these was headed 'Piston-Engined Aircraft' and stated that all surfaces, except the upper surface of the fuselage was to be coloured Dark Sea Grey BS 381C/638. The upper surface of the fuselage was to be white and on Near East Air Force aircraft a similar white finish was to be applied over the area of the wing fuel tanks. The usual national markings were to be carried on the upper and under surfaces of the mainplane, both sides of the fuselage and on each side of the fins in Post Office Red BS 381C/538, White and Roundel Blue BS 381C/110, whilst the aircraft serial numbers and identification letters were to be applied in Post Office Red edged in white.

Most of these markings were in evidence on 8 Squadron's Shackleton AEW.2s in June 1972 including the new underwing roundels of 36 inch diameter by which time the Squadron colours were repositioned to flank the roundel 'fighter style' leaving the Squadron badge on the nose. The fuselage serial was applied using Post Office Red only and in place of an individual aircraft letter, the 'last two' of the serial number were applied in Post Office Red and white above the flash on the outer faces of the fins. In contrast to the Shackleton fleet of earlier years, the AEW.2 did not have the heat reflecting gloss White upper surface on the fuselage. The requirement for the heat reflecting white upper surface on the fuselage and wings was absent from DCI S.68 of 23 April 1975 which was otherwise unchanged from DCI S.70/72.

Between 1972 and 1974, 8 Squadron's AEW.2s carried 'nose art' in the form of an illustration of one of the characters from the then well-known children's television series 'Magic Roundabout' and 'The Herbs', which was painted beneath the side window of the pilot's cockpit on the port side of the fuselage.

In March 1989 the Squadron number reappeared in Post Office Red and white on the fuselage side which then remained until the Shackleton AEW.2 finally retired in 1991.

South African Air Force

The eight South African Air Force (SAAF) MR.3s (serial numbers 1716 to 1723) were delivered with Dark Sea Grey upper surfaces and PRU Blue under surfaces with the demarcation between the two slightly lower down the side of the fuselage than had been the case on most RAF Shackletons. In the case of the SAAF Shackletons, the demarcation coincided with the bottom edge of the main cockpit canopy.

The first three Shackletons delivered, 1716, 1717 and 1719, carried a SAAF roundel in blue and white with an orange springbok at its centre in all six positions with the springbok facing inboard on the upper and underwing roundels and forward on the fuselage roundels. These were of a similar size to the roundels applied to RAF Shackletons. Subsequent aircraft were marked with the revised SAAF 'castle' insignia of approximately 48 inch diameter in blue with a white outline with an orange springbok once again at its centre facing in the same directions as previous roundels. All the Shackletons carried an RAF-style, equally proportioned, fin flash in orange, white and blue with orange leading on the outer faces of the fins only.

The serial number was applied in black approximately 24 inches high to both sides of the rear fuselage just forward of the tailplane and approximately 48 inches high under the mainplanes inboard of the national marking in a similar manner to RAF serial numbers in so far as the serial number could be read from behind the aircraft on the port wing and from the front of the aircraft on the starboard wing.

In SAAF service, the Shackletons were operated by 35 Squadron which employed a yellow individual aircraft letter approximately 48 inches high in the range 'J' to 'Q' inclusive, just aft of the astrodome. Known code allocations were 1716 'M', 1717 'O', 1718 'K', 1720 'M', 1721 'N' and 1722 'P'. In addition to these markings, the Squadron badge was applied to the forward fuselage. Initially, the Squadron badge of a pelican astride a globe on waves showing the African continent, with the motto *Haya Amanzi* (Strike the Water), was applied in a shield but this was later changed to a circular design. The last two digits of the serial number were applied on the nose between the guns in black.

Following delivery, the original scheme was modified by the addition of a white finish to the upper surface to the fuselage and a black anti-glare panel to the nose, similar to the RAF machines. The addition of the white upper surface to the fuselage left a narrow strip of Dark Sea Grey running the length of the fuselage. This was eliminated during 1969 by extending the PRU Blue up the side of the fuselage to meet the bottom edge of the white fuselage top. In addition to these changes, the spinners were painted red.

South African Air Force Shackletons were initially delivered with Dark Sea Grey upper surfaces and PRU Blue under surfaces with SAAF roundels in all six positions, although the aircraft were subsequently re-marked with the revised SAAF 'castle' insignia, as illustrated by 1716 coded 'J' in yellow. Following delivery, the original scheme was modified by the addition of white to the upper surface of the fuselage and a black anti-glare panel to the nose in a similar manner to RAF machines. The spinners were painted red. *Author's collection*

Appendices

■ APPENDIX 1   **SQUADRONS AND OTHER MAJOR UNITS**

NOTES

Variants used – provides a guide to the primary Mark(s) operated with dates.

Period used – refers to the overall period in which Shackletons were used, irrespective of Mark.

Primary bases – is self-explanatory; deployments or detachments are not shown.

It shouldn't be assumed that each unit reformed on the dates shown as some simply re-equipped with Shackletons. Equally, at the other end of the scale, it cannot be assumed that each automatically disbanded on the last date shown as, in some instances, new types were received to replace the old.

No.8 Squadron
Variant(s) used	AEW.2 (plus small number of MR.2 crew trainers)
Period used	1/4/72 - 30/6/91
Primary base(s)	Kinloss (1/72 - 8/73), Lossiemouth (8/73 - 6/91)

No.37 Squadron
Variant(s) used	MR.2
Period used	7/53 - 7/9/67
Primary base(s)	Luqa, Malta (7/53 - 7/57), Aden (7/57 - 9/67)

No.38 Squadron
Variant(s) used	MR.2
Period used	18/9/53 - 31/3/67
Primary base(s)	Luqa, Malta (9/53 - 9/65), Hal Far, Malta (9/65 - 3/67)

No.42 Squadron
Variant(s) used	MR.1/1A (6/52-7/54), MR.2 (1/53-6/66), MR.3 (11/65-9/71)
Period used	28/6/52 - 9/71
Primary base(s)	St Eval (6/52 - 10/58), St Mawgan (10/58 - 9/71)

No.120 Squadron
Variant(s) used	MR.1/1A (3/51-10/56), MR.2 (3/53-8/54 & 10/56-12/58), MR.3 (11/58-2/71)
Period used	3/51 - 2/71
Primary base(s)	Kinloss (3/51 - 4/52), Aldergrove (4/52 - 4/59), Kinloss (4/59 - 2/71)

No.201 Squadron
Variant(s) used	MR.3
Period used	1/10/58 - 12/70
Primary base(s)	St Mawgan (10/58 - 7/65), Kinloss (7/65 - 12/70)

No.203 Squadron
Variant(s) used	MR.1A (11/58-2/59), MR.2 (4/62-12/66) MR.3 (11/58-7/62 & 6/66-12/71)
Period used	1/11/58 - 12/71
Primary base(s)	Ballykelly (11/58 - 2/69), Luqa (2/69 - 12/71)
Note	This sqn didn't operate the MR.1

No.204 Squadron
Variant(s) used	MR.1/1A (5/58-2/60), MR.2 (1/54-5/58 & 6/59-4/72)
Period used	1/1/54 - 4/72
Primary base(s)	Ballykelly (1/54 - 4/71), Honington (1/4/71 - 28/4/72)
Note	Further information regarding this unit appears in caption for MR.2, WG555 on page 20

No.205 Squadron
Variant(s) used	MR.1/1A (5/58-9/62), MR.2 (2/62-10/71)
Period used	5/58 - 10/71
Primary base(s)	Changi (5/58 - 10/71)

No.206 Squadron
Variant(s) used	MR.1/1A (9/52-6/58), MR.2 (2/53-6/54), MR.3 (1/58-10/70)
Period used	27/9/52 - 10/70
Primary base(s)	St Eval (9/52 - 1/58), St Mawgan (1/58 - 7/65), Kinloss (7/65 - 10/70)

No.210 Squadron
Variant(s) used	MR.2
Period used	1/12/58 - 12/71
Primary base(s)	Ballykelly (12/58 - 10/70)
Note	Disbanded at Ballykelly on 31/10/70. 210 Sqn reformed next day at Sharjah, with MR.2s, until disbanded there on 15/11/71

No.220 Squadron
Variant(s) used	MR.1/1A (9/51-2/58), MR.2 (2/53-7/54), MR.3 (8/57-10/58)
Period used	24/9/51 - 1/10/58
Primary base(s)	Kinloss (9/51 - 11/51), St Eval (11/51 - 12/56), St Mawgan (12/56 - 10/58)
Note	Disbanded 1/10/58 and renumbered 201 Squadron operating the MR.3

No.224 Squadron
Variant(s) used	MR.1/1A (7/51-9/54), MR.2 (5/53-10/66)
Period used	7/51 - 31/10/66
Primary base(s)	Gibraltar (7/51 - 10/66)

No.228 Squadron
Variant(s) used   MR.2
Period used   1/7/54 - 6/3/59
Primary base(s)   St Eval (7/54 - 11/56),
St Mawgan (11/56 - 1/58),
St Eval (1/58 - 3/59)

No.240 Squadron
Variant(s) used   MR.1/1A (5/52-11/58),
MR.2 (3/53-8/54)
Period used   1/5/52 - 1/11/58
Primary base(s)   Aldergrove (5/52 - 5/52),
St Eval (5/52 - 6/52),
Ballykelly (6/52 - 11/58)
Note   Disbanded 1/11/58 and renumbered
203 Squadron operating MR.1A then
MR.2

No.269 Squadron
Variant(s) used   MR.1/1A (1/52-11/58),
MR.2 (3/53-8/54 & 10/58-12/58)
Period used   1/1/52 - 1/12/58
Primary base(s)   Gibraltar (1/52 - 3/52),
Ballykelly (3/52 - 12/58)
Note   Disbanded 1/12/58 and renumbered
as 210 Squadron operating the MR.2

No.236 Operational Conversion Unit
Variant(s) used   MR.1/1A
Period used   31/5/51 - 1/10/56
Primary base(s)   Kinloss (5/51 - 10/56)
Note   236 OCU and School of Maritime
Reconnaissance combined on
1/10/56 to form Maritime
Operational Training Unit (MOTU).
At least one MR.2, WR966, was used
briefly by this unit in October 1970 –
MOTU having reverted to 236 OCU
on 1/7/70

Maritime Operational Training Unit
Variant(s) used   MR.1/1A, T.4, MR.2
Period used   1/10/56 - 1/7/70
Primary base(s)   Kinloss (10/56 - 7/7/65),
St Mawgan (7/65 - 7/70)
Note   MOTU reverted to 236 OCU at St
Mawgan 1/7/70

Joint Anti-Submarine School (JASS or JASS Flight)
Variant(s) used   MR.1A, MR.2
Period used   18/3/52 - 3/57
Primary base(s)   Ballykelly (3/52 - 3/57)
Note   Flight established at three Mk.1A,
replaced by three MR.2s in 1955. At
least four MR.1A and four MR.2 were
operated in total

Air-Sea Warfare Development Unit
Variant(s) used   MR.1/1A, T.4, MR.2, MR.3
Period used   4/51 - 1/4/70
Primary base(s)   St Mawgan (3/52 - 1/9/58)
Ballykelly (1/9/58 - 4/70)

School of Maritime Reconnaissance (SMR), 1 Maritime Reconnaissance School (MRS)
Several sources suggest that one or more MR.1s might have been used by these units prior to MOTU forming, but no evidence has come to light to support this. It is suggested that a T.4 was also allocated, but that's not possible as the first T.4 conversions had yet to be completed in October 1956 when MOTU was formed.

Shackletons were also allocated to such as the RAE at Farnborough, and A&AEE at Boscombe Down, some of which are included in the photograph section. MR.1 VP256, and MR.2 WG533 were used briefly by the RAF Flying College, Manby, to produce the Marks' pilot notes.

In comprehensive Shackleton histories, in addition to Maintenance Units, readers might occasionally find references to the Coastal Command Aircraft Preparation & Modification Flight which formed on 12/3/51 at Valley, prior to moving to St Mawgan two months later. Redseignated Coastal Command Modification Centre (CCMF) on 15/7/54, it disbanded on 31/12/55. A similar unit, the Coastal Command Preparation & Modification Unit was formed at Ballykelly on 12/5/51. It too later moved to St Mawgan and was subsequently absorbed by the CCMF. These units did not operate aircraft, their purpose was to modify Shackletons (and Lancasters) as necessary for the maritime role.

SOUTH AFRICAN AIR FORCE

No.35 Squadron
Variant(s) used   MR.3
Period used   1967 - 1984
Primary base(s)   D.F. Malan Airfield, Cape Town.

Issued originally to 224 Squadron in January 1952 as an MR.1A coded 'B-K', WB832 is seen here *circa* 1962 as T.4 'U' belonging to MOTU. Converted at Langar in 1956 this aircraft had earlier been operated by MOTU (coded 'G') prior to being modified between January and October 1961. Upon its return WB832 was subsequently recoded 'U'. *Newark Air Museum*

■ APPENDIX 2 **CODE LETTER ALLOCATION** – 1951-1954

Ballykelly
204 Squadron code T: Aircraft codes R to Y
240 Squadron code L: Aircraft codes A to G and T
269 Squadron code B: Aircraft codes A to H and J
JASS Flight unit code G: Aircraft codes W to Y (For air-to-air identification JASS Shackletons had twin black bands applied to their wings with a single, interrupted, broad band around the fuselage.)

Gibraltar
224 Squadron code B: Aircraft codes A to H and J (replaced by K to T)

Kinloss
120 Squadron code A: Aircraft codes A to H and J
236 OCU code C: Aircraft codes H to Z

Luqa
37 Squadron aircraft codes A to H
38 Squadron aircraft codes Q and S to Z
Neither squadron employed a unit code, they simply applied their individual aircraft letters aft of the fuselage roundel

St Eval
 42 Squadron code A: Aircraft codes A to H and J
206 Squadron code B: Aircraft codes T to Z
220 Squadron code T: Aircraft codes K to S
228 Squadron code L: Aircraft codes K to U

St Mawgan
ASWDU code F: Aircraft codes A, B, D, F and K

■ APPENDIX 3 **INDIVIDUAL CODE LETTER ALLOCATION** – 1962

Aden
37 Squadron aircraft codes A to D

Ballykelly
203 Squadron: E to H, J and K
204 Squadron: M to R
210 Squadron: T to Z
ASWDU: A and B
Station Flight: C

Gibraltar
224 Squadron: A to C, P to Z

Kinloss
120 Squadron: A to G
MOTU: G to Z

Luqa
38 Squadron: S to Z

St Mawgan
42 Squadron: A to F
201 Squadron: K to P
206 Squadron: A to F

Changi, Singapore
205 Squadron: A to H, J to N

■ APPENDIX 4 **BASIC SPECIFICATIONS**

	MR.1/1A T.4	MR.2/AEW.2	MR.3
Wing span	120ft 0in (36.57m)	120ft 0in (36.57m)	119ft 10in (39.29m)
Length	77ft 6in (23.62m)	87ft 4in (26.61m)	87ft 4in (26.61m)
Tailplane span	33ft 0in (10.06m)	33ft 0in (10.06m)	33ft 0in (10.06m)

Propellers — All Marks: De Havilland 13ft diameter counter-rotating units

Guns — Production MR.1 and MR.1As were initially armed with two 20mm Hispano cannon mounted in a Bristol B.17 dorsal turret
MR.2s initially received four 20mm Hispano cannon – two each in nose position and dorsal turret
MR.3s initially carried twin 20mm Hispano cannon in the nose only
(The T.4, T.2 and AEW.2 were unarmed)

MR.2 prototype WB833. For further information relating to this aircraft see pages 19 and 69.
Author's collection

■ APPENDIX 5 **SERIAL NUMBER ALLOCATIONS**

Shackleton MR.1 & 1A

Prototypes	VW126, VW131, VW135	(3 completed)
MR.1 production	VP254-VP268 & VP281-VP294	(29 completed – VP253 cancelled when prototypes were ordered)
MR.1A production	WB818-WB837 & WB844-WB861	(37 completed as MR.1A, plus WB833 built as MR.2 prototype)
MR.1A production	WG507-WG511 & WG525-WG529	(10 completed as MR.1A – remainder of order, WG530-533, WG553-558 as MR.2)
T.4 conversions	VP258, VP259, VP293, WB819, WB820, WB822, WB826, WB831, WB832, WB837, WB844, WB845, WB847, WB849, WB858, WG511 and WG527.	(Note: WB854's conversion was started and then abandoned)

Shackleton MR.2

MR.2 Prototype	WB833	(1 completed)
MR.2 Production	WG530-WG533, WG553-WG558, WL737-WL759, WL785-WL801, WR951-WR969, WR970-WR990	(69 units completed – remainder of original order, completed as MR.3)

Shackleton MR.3

MR.3 Prototype	WR970	(1 completed)
MR.3 Production	WR971-WR990, XF700-XF711 plus XF730	(33 completed)
For SAAF	1716-1723	(8 delivered)

■ APPENDIX 6 **KNOWN AIRCRAFT LOSSES** – BY DATE

Type	Serial	Date	Unit	Notes
MR.1	VP283	12/8/51	224 Sqn	Crashed into sea on approach to runway at Gibraltar
MR.1	VP261	25/6/52	120 Sqn	Crashed into sea 12 miles off Berwick on Tweed. 10 crew killed
MR.1	VP286	8/10/52	239OCU	Crashed into sea at night, Tarbet Ness, cause not known. 14 killed
MR.2	WL749	14/5/53	120 Sqn	Undershot runway at Aldergrove
MR.2	WL746	11/12/53	240 Sqn	Crashed into sea off Hebrides, cause not known. All 10 crewmen killed
MR.2	WL794	12/2/54	38 Sqn	Flew into sea near Gozo, Malta, during ASW exercise. 10 crewmen killed
MR.1	VP256	26/10/54	269 Sqn	Overshot runway at Ballykelly. Aircraft written off
MR.2	WG531	11/1/55	42 Sqn	Believed to have collided with WL743 south of Fastnet Rock. 9 killed
MR.2	WL743	11/1/55	42 Sqn	Believed to have collided with WG531 south of Fastnet Rock. 9 killed
MR.2	WL799	22/12/55	38 Sqn	Undergoing mods at Langar, a/c caught fire and was destroyed
MR.3	WR970	7/12/56		Undergoing stalling tests prototype MR.3 crashed in Derbyshire. 4 killed
MR.1A	WB861	5/9/57	240 Sqn	Made a wheels-up landing and was SOC at Ballykelly
MR.2	WL792	14/9/57	224 Sqn	Wheels-up landing at Gibraltar following loss of power – a/c scrapped
T.4	VP259	10/1/58	MOTU	Struck trees in low cloud, hit ground and caught fire. 2 dead with 7 survivors
MR.1	VP254	9/12/58	205 Sqn	Believed a/c flew into South China Sea while on low-level patrol. 11 dead
MR.2	WR968	20/10/61	210 Sqn	Crashed while landing at Ballykelly, a/c burnt out
MR.1	VP294	15/5/62	205 Sqn	Force-landed at Gan and damaged beyond repair
MR.3	XF710	10/1/64	120 Sqn	Crashed on Culloden Moor following an uncontrollable fire in stbd wing
MR.3	XF704	8/12/65	201 Sqn	Crashed into the sea on approach to RAF Kinloss. 7 killed
MR.2	WL786	4/11/67	205 Sqn	Ditched in Indian Ocean en route to Gan. 8 killed, 3 survivors
MR.3	WR976	19/11/67	201 Sqn	A/c struck the sea during exercise near Scilly Isles. 9 killed, 2 survivors
MR.3	XF702	21/12/67	206 Sqn	A/c dived into the ground in severe weather near Inverness. 11 killed
MR.2	WR956	1/4/68	204 Sqn	Crash-landed at Ballykelly and a/c SOC
MR.2	WB833	19/4/68	210 Sqn	Crashed into ground on Mull of Kintyre following ASW exercise. 11 dead
AEW.2	WR965	30/4/90	8 Sqn	Crashed into mountain on Isle of Harris killing all 10 crewmen

Avro Shackleton MR.1, VW131, second prototype, displayed at the SBAC show Farnborough, Hampshire, in September 1949 | VW131 was displayed at the SBAC show finished in the then current overall White with Medium Sea Grey upper surfaces Coastal Command scheme, with post-war national markings and the yellow prototype 'P' marking within a circle on the rear fuselage. The aircraft was fitted with the originally specified nose barbettes – each with a single 20mm Hispano cannon, two 20mm cannon in a Bristol B.17 dorsal turret, and two .5in mgs mounted in a Boulton Paul Type D tail turret. VW131 was used in a series of tests until it was stored. Dismantled in 1956, the fuselage was retained for ditching trials until 1962.

Shackleton MR.1, VP258 'A-C' of 120 Squadron, RAF Aldergrove, Northern Ireland, 1953 | VP258 was delivered to 120 Squadron on 3 April 1951 where it received the Light Slate Grey code 'A-C' placed either side of the fuselage roundel – 'A' being the Squadron code and 'C' the individual aircraft letter, which was repeated on the nose. Finished in the standard Coastal Command overall White with Medium Sea Grey upper surfaces scheme, with post-war national markings, this aircraft remained with the unit until May 1955 after which it was sent to Woodford where it was converted to become the prototype Shackleton T.4. VP258 displays 120 Squadron's badge within a standard frame on its nose illustrating an Icelandic falcon perched on a globe.

Shackleton MR.1A, WB851 'G-Y' of the Joint Anti-Submarine School (JASS) Flight, RAF Ballykelly, County Derry, Northern Ireland, 1952 | The Joint Anti-Submarine School (JASS) Flight was manned by both RAF and RN personnel for the development and practice of anti-submarine tactics. Finished in the then standard Coastal Command overall White with Medium Sea Grey upper surfaces scheme with post-war national markings, its Shackletons were distinctively marked with twin black bands around their outer wings and a single, broader, band around the mid-fuselage flanking the roundel. WB851 is illustrated as it appeared in 1952 wearing Light Slate Grey codes. WB851 later served with several squadrons prior to joining ASWDU in November 1959.

Shackleton MR.1A, WB834 'C-L' of 236 OCU, RAF Kinloss, Moray Firth, Scotland, 1956 | WB834 is illustrated finished in the overall high gloss Dark Sea Grey scheme introduced in mid-1955. The code 'C', identifies 236 OCU while 'L' is the individual aircraft letter. The codes and serial numbers, both on the fuselage and under the wings, were initially all in white, although they later changed to red with the introduction of the revised identification markings in 1956. WB834 went on to serve with the Maritime Operational Training Unit (MOTU) when that unit formed in late 1956, and remained with the latter until August 1957 when it was modified and stored. Allocated to 205 Squadron in the Far East in late 1958, WB834 was placed in storage at Singapore in April 1961 and SOC four months later.

These scrap views illustrate different underwing serial presentations on sample airframes:
A: The original black serials | **B:** White serials initially introduced with the overall high gloss Dark Sea Grey scheme in mid-1955 |
C: The red serials outlined in white introduced in 1956 and which remained in use throughout the remainder of the Shackleton's operational service, in this instance, under the wings of WB850.

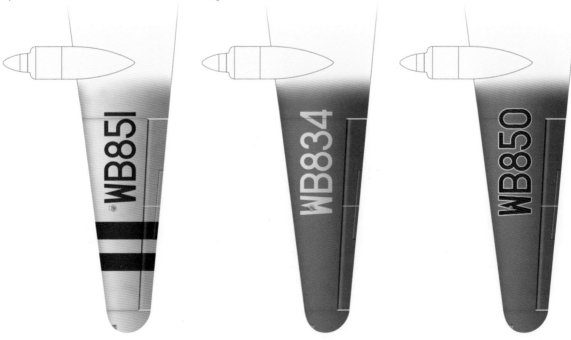

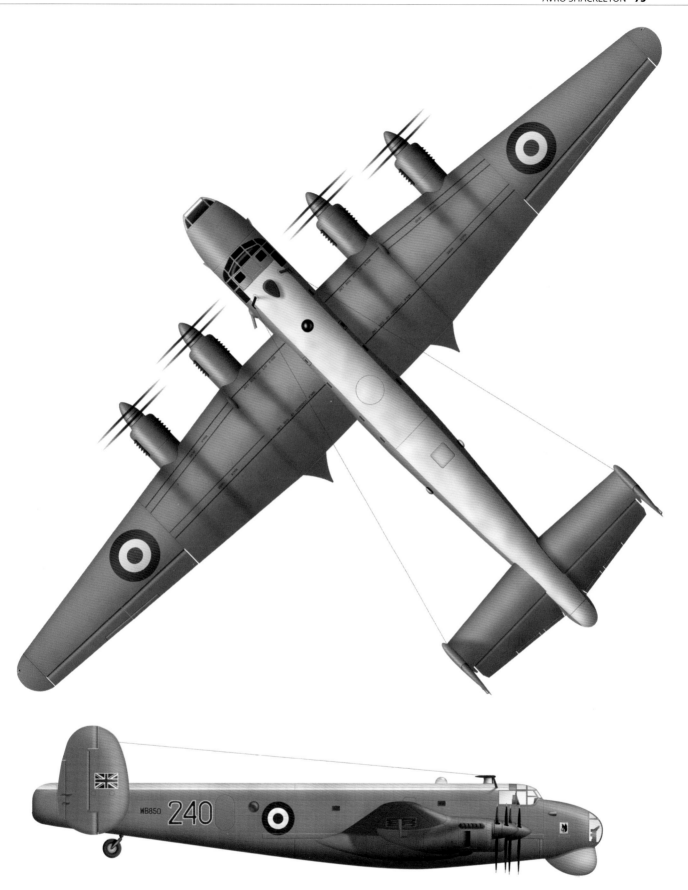

Shackleton MR.1, WB850 of 240 Squadron, RAF Ballykelly circa 1957 | WB850 is illustrated as it appeared following the introduction of the revised identification markings which dispensed with unit identification codes in lieu of a squadron number positioned on the rear fuselage in large red numerals outlined in white. Finished in the overall Dark Sea Grey scheme, the upper fuselage was painted white for overseas detachment and the fin flashes on the outer faces of the fins were replaced by Union Flags – deemed more distinctive when 'showing the flag' abroad. Roundels were applied above the wings and on the fuselage sides, with red serials on the rear fuselage and under the wings – the latter outlined in white. WB850 carried the Squadron's winged Icelandic helmet motif and the name *Mjölner* (the name of Thor's hammer) on the nose, commemorating the time spent by the unit in Iceland in WWII. WB850 was transferred to 204 Squadron in June 1958.

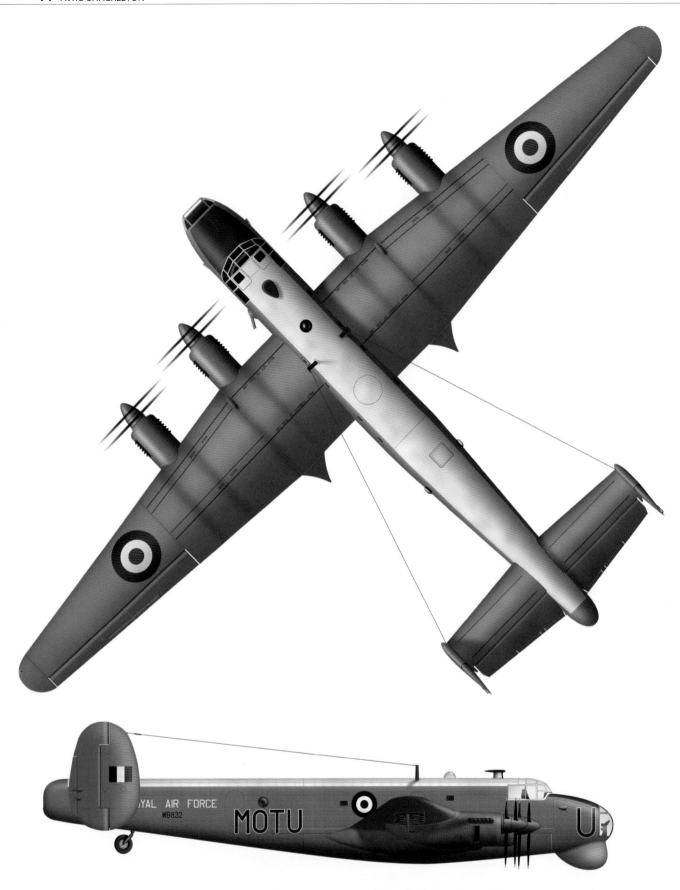

Shackleton T.4, WB832 belonging to MOTU, RAF Kinloss, circa 1962 | Built as an MR.1A, WB832 was subsequently converted to T.4 standard and allocated to MOTU, coded 'G'. After being further modified, between January and October 1961, WB832 returned to the unit, albeit re-coded 'U' as illustrated here, wearing the overall Dark Sea Grey scheme with white applied to the upper fuselage which stopped just short of the end of the rear fuselage. In lieu of a squadron number, the letters 'MOTU' were applied to the mid-fuselage position, in red outlined in white, with the similarly coloured individual aircraft letter 'U' on the nose. At the end of 1961, the legend 'ROYAL AIR FORCE' was being applied to the rear fuselage in white, above the red serial number, and the fuselage roundel was re-positioned further forward and higher. Following its withdrawal from frontline use in July 1965, WB832 was sent to RAF Cosford to become 7885M and was finally scrapped there in 1974.

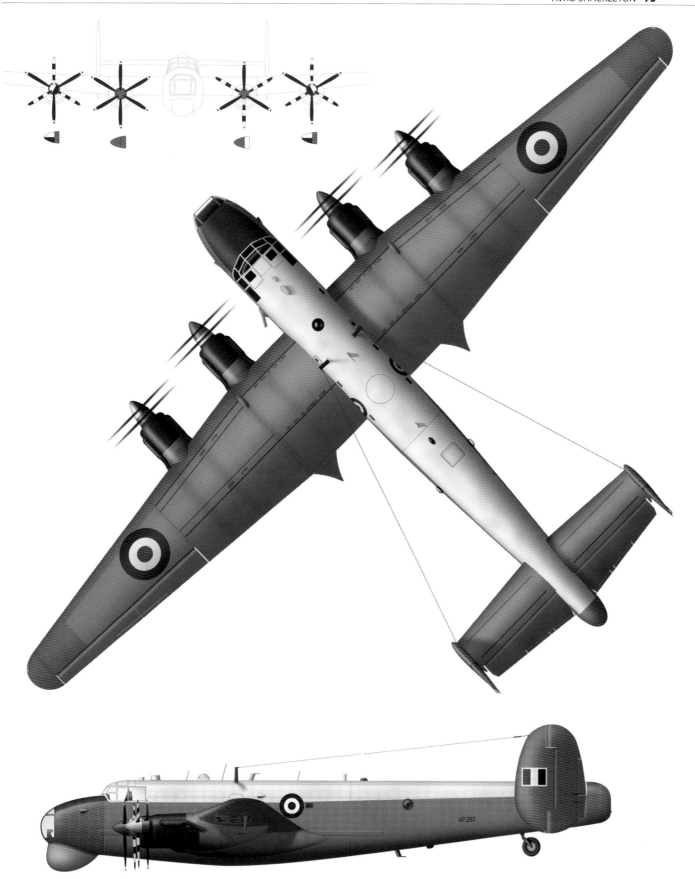

Shackleton T.4, VP293 of the Weapons Flight, Royal Aircraft Establishment (RAE) Farnborough, Hampshire, 1973 | Built as an MR.1, VP293 first flew in July 1951 and eventually became the longest-lived MR.1/T.4. Conversion to T.4 standard commenced in 1956 with VP293 ultimately being received by MOTU following a period of trials with the A&AEE. Following storage at 23 MU in early 1963, VP293 was selected for use by the RAE at Farnborough in January 1964. Finished in the standard RAF Coastal Command scheme of overall gloss Dark Sea Grey with a white upper fuselage demarcation that extended lower than usual, the nose, fins, extreme rear fuselage and wing leading edges and wingtips were painted in Dayglo orange. The engine cowlings forward of the wing leading edges were painted black at this stage, with the propeller spinners in differing combinations of black, white and Dayglo orange, and the propeller blades in combinations of black and white stripes or black with red and white tips (see scrap view). A black anti-glare panel was applied and, reportedly, a cartoon 'Zebedee' featured on the end of the fuselage after the fictional 'Magic Roundabout' TV character – inspired by much of the internal equipment having to be mounted on springs to dampen vibration.

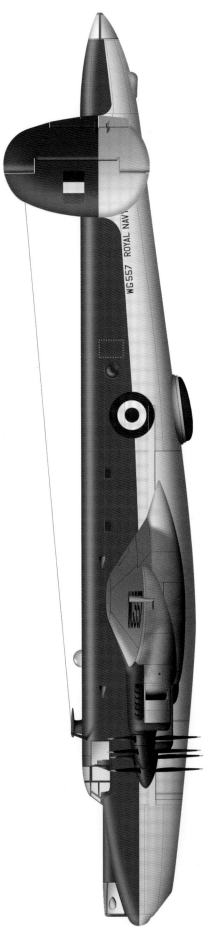

Shackleton MR.2, WG557 'T-L' of 220 Squadron, RAF St Eval, Cornwall, mid-1954 | WG557 was initially allocated to 220 Squadron in March 1954, but was transferred to 228 Squadron six months later. Illustrated here while serving with 220 Squadron, it was finished in the then standard overall White with Medium Sea Grey upper surfaces scheme with Light Slate Grey codes and fuselage serial numbers, with those under the wings in black. In December 1957, WG557 was allocated to the RAE at Farnborough for use in the development of fusing systems for nuclear weapons as well as other duties.

Shackleton MR.2, WG557, Empire Test Pilots School, Farnborough, Hampshire, September 1964 | At some point in its service with the ETPS, WG557 was repainted in a unique scheme comprising what appears to be a medium blue shade on the upper surfaces, with a much lower demarcation line along the fuselage. The upper half of the fins were also painted in the same shade. Fuselage serials were repainted in red, as were the underwing serials, and a black 'ROYAL NAVY' legend was added to the rear fuselage. WG557 ended its days relegated to Farnborough's dump.

Shackleton MR.2, WR963 'B-M', of 224 Squadron, Gibraltar, 1954 | Finished in the overall White with Medium Sea Grey upper surfaces scheme, both code letters were applied in black, as were the underwing serial numbers, while the fuselage serial numbers were Light Slate Grey. The unit's serpent-entwined rock motif was carried on the outer face of both fins on a white shield. WR963 later went on to become an AEW.2 with 8 Squadron at Lossiemouth.

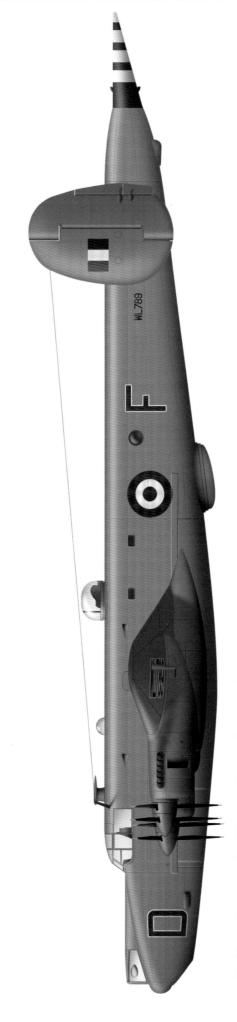

Shackleton MR.2, WL789 'F-D', Air-Sea Warfare Development Unit, RAF St Mawgan, Cornwall, 1953 | Tasked with the development and testing of new maritime equipment, the ASWDU tended to change its aircraft depending on what trials were being conducted, and during this time examples of both early Marks of Shackleton were on the unit's strength. Finished in the overall Dark Sea Grey scheme, WL789 was fitted with a magnetic anomaly detector (MAD) in an extended tail boom, which it is thought was painted with blue and white bands, (as illustrated), but which may just have been white bands around the standard Dark Sea Grey body. The code letters 'F' and 'D' were in the standard red outlined in white. Despite continued attempts to resolve the matter, MAD proved incompatible with the Shackleton and was not adopted.

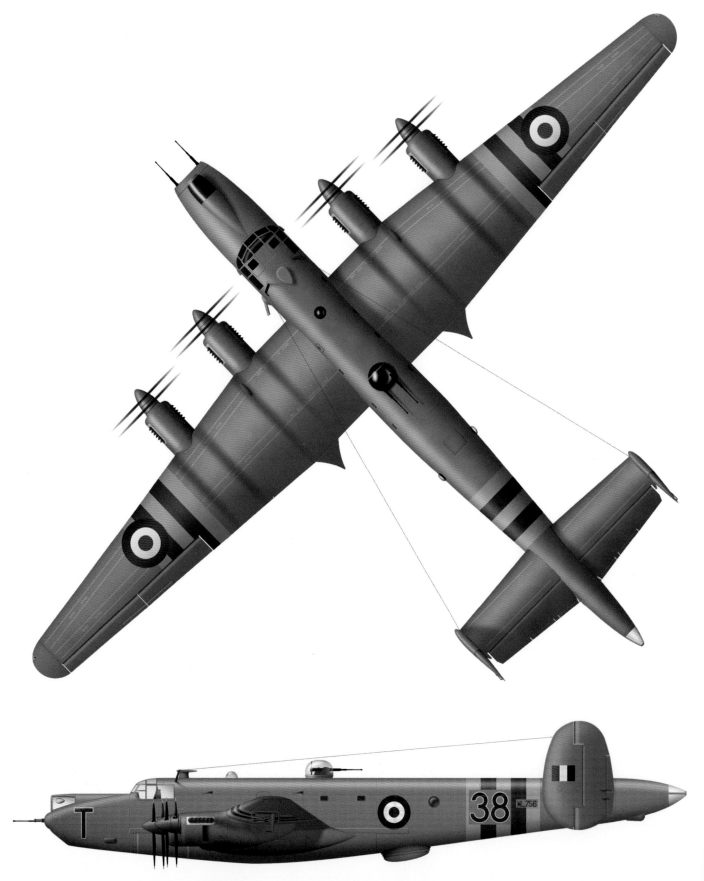

Shackleton MR.2, WL756 'T' of 38 Squadron, Luqa, Malta, November 1956 | As one of the RAF squadrons involved in Operation *Musketeer*, the invasion of the Suez Canal area of Egypt in November 1956, 38 Squadron's Shackleton MR.2s had the yellow and black 'distinctive markings' applied which identified allied aircraft during the Operation. WL756 was finished in the overall Dark Sea Grey scheme and carried the Squadron number (introduced on Coastal Command aircraft in March 1956) on the rear fuselage, with the individual aircraft letter remaining on the nose, both in red outlined in white. Note how the 'Suez Stripes', consisting of three yellow and two black stripes, each 24 inches wide, were painted around the Squadron number, the fuselage and underwing serials and were not applied over the ailerons. Shackletons of both 37 and 38 Squadrons, operating from Malta, adopted these markings with effect from D minus 1.

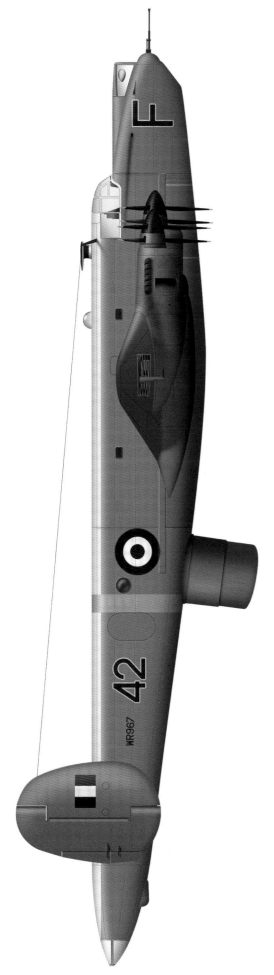

Shackleton MR.2, WR967 'F', of 42 Squadron, RAF St Eval, Cornwall, 1957 | Reformed in June 1952, 42 Squadron received its first MR.2s in January 1953 and despite being ostensibly based at St Eval, undertook numerous detachments to the Mediterranean and Middle East, for which its aircraft had the upper surface of their fuselages painted in glossy white to reflect some of the heat. Otherwise finished in the standard overall Dark Sea Grey scheme, WR967, coded 'F' and sporting the Squadron number, featured a yellow stripe around the mid-fuselage just to the rear of the radome, which may have been an exercise marking. Note the red spinners.

Shackleton MR.2, WL796, Farnborough, Hampshire, fitted with a Mk.3 Lifeboat, September 1953 | WL796 is illustrated as it looked after being modified to carry the Mk.3 Lifeboat which it demonstrated at Farnborough in September 1953, although in the event the lifeboat was never actually dropped from an MR.2 due to the introduction of Lindholme Gear, (an air-dropped self-inflating dinghy). Finished in the then standard Medium Sea Grey and White scheme, WL796 carried a 'Hawker Siddeley Group' logo on its nose while on display at Farnborough.

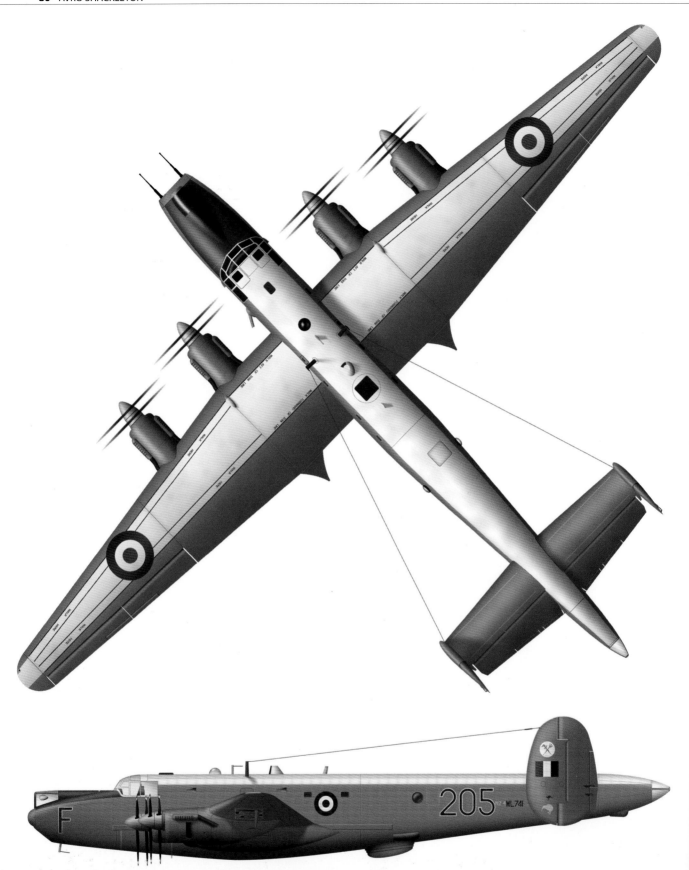

Shackleton MR.2, WL741 'F' of 205 Squadron, RAF Changi, Malaya, 1967 | Part of the Far East Air Force, 205 Squadron's MR.2s featured heat-reflecting glossy-white fuselage upper surfaces from the beginning, as illustrated on WL741, with a similar glossy white spanwise strip over the area of the wings' fuel tanks, which in most instances extended to the tips – but not in every case. Otherwise finished in the standard overall Dark Sea Grey scheme it had the Squadron number '205' on the rear fuselage and the individual aircraft letter 'F' on the nose, both in red outlined in white. The Squadron badge, a kris and a trident crossed on a white disc, was carried on the outer faces of both fins. Evidence of 'zaps' are apparent – two from New Zealand (a yellow kiwi silhouette on the fin and a yellow 'NZ4' prefix to the fuselage serial) plus one from Australia (a black kangaroo silhouette on the fin above the kiwi), indicative of a recent joint, or combined, exercise perhaps?

Shackleton AEW.2 development aircraft, WL745 'O', RAE Farnborough, Hampshire, September 1971 | WL745 was used to test the type's compatibility with the AN/APS-20 airborne early warning radar system and effectively became the prototype Shackleton AEW.2 albeit finished in the overall Dark Sea Grey scheme with a white upper fuselage and black anti-glare panel in front of the cockpit. WL745 was fitted with a large radome under the forward fuselage and a temporary probe in the nose, while the ASV radar housing was removed and blanked-off. Following completion of the trial phase, WL745 was upgraded to production standard and issued to 8 Squadron in September 1973, becoming the last of twelve AEW.2s to join the unit.

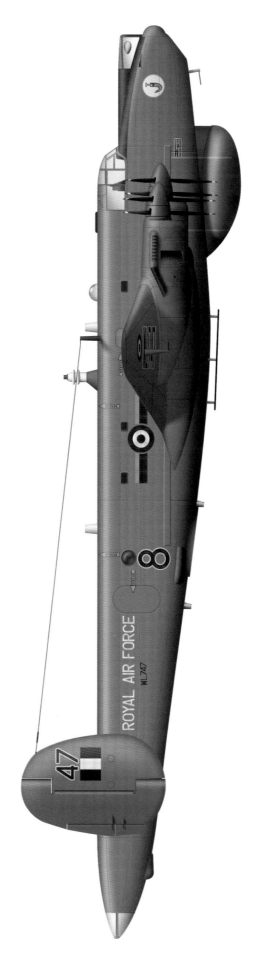

Shackleton AEW.2, WL747 '47', of 8 Squadron, RAF Lossiemouth, August 1988 | Number 8 Squadron reformed at RAF Kinloss on 1 January 1972 undertaking the RAF's Airborne Early Warning (AEW) role equipped with Shackleton AEW.2s, the only unit to operate the type. It moved to RAF Lossiemouth in August 1973, where it stayed until its disbandment in 1991. WL747 is illustrated in the standard AEW finish of overall gloss Dark Sea Grey, but in contrast to the rest of the Shackleton fleet, AEW.2s did not have heat-reflecting gloss white fuselage upper surfaces. The usual national markings were carried including underwing roundels and a white 'ROYAL AIR FORCE' legend was positioned above the red fuselage serial number. Squadron colours, comprising sand, blue and red bars, were carried 'Fighter style' either side of the fuselage roundel with the Squadron badge, an Arabian dagger on a white disc within a pale blue surround, on the nose. The Squadron number on the fuselage side, the 'last two' of the serial number on the outer faces of the fins, and the underwing serials were red outlined in white.

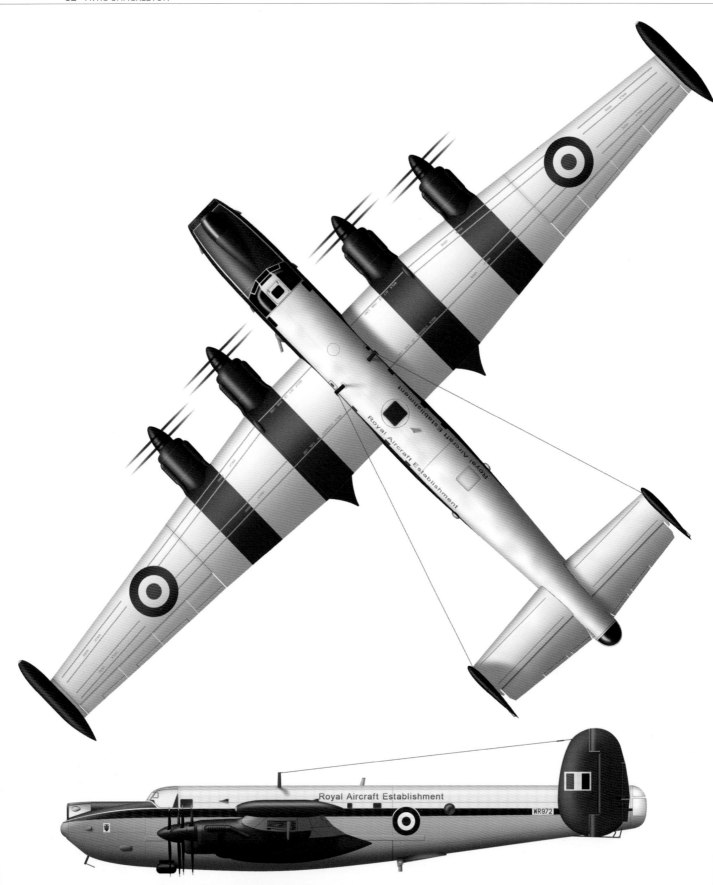

Shackleton MR.3, WR972, Structures and Mechanical Engineering Flight, RAE Farnborough, summer 1971 | Illustrated as it looked while serving with the RAE in the late 1960s/early 1970s, WR972 was finished in a Light Aircraft Grey and white scheme, with the upper surfaces of the wings, tailplanes and fuselage in white, with a Roundel Blue cheatline separating the Light Aircraft Grey and the white, running the full length of the fuselage. Both faces of the fins and rudders, tip tanks and propeller spinners were also Roundel Blue. Serials and engine nacelles were painted black, with black 'exhaust-masking' chordwise bands over the upper surfaces of the wings. The 'Royal Aircraft Establishment' title in black was applied to the fuselage sides and the RAE motif, on a white rectangular background, was carried on the nose below the cheatline and black anti-glare panel. WR972 ended its days in February 1973 on the Farnborough fire dump.

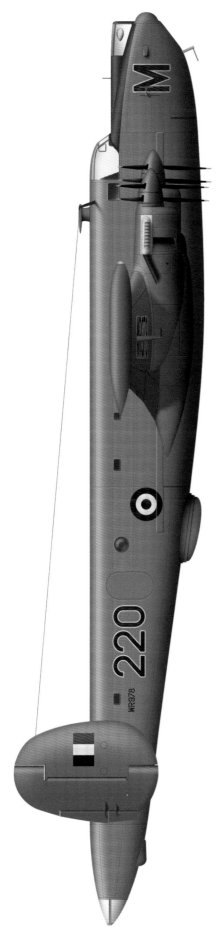

Shackleton MR.3, WR978 'M' of 220 Squadron, RAF St Mawgan early 1958 | Finished in the standard overall gloss Dark Sea Grey scheme, without a white upper fuselage, presumably because this unit was not scheduled to be sent on overseas detachments. The Squadron number, individual aircraft letter and underwing serials were all red outlined in white, with just the fuselage serial in plain red. Number 220 Squadron only used the MR.3 for a short period, from August 1957 to October 1958, when it became 201 Squadron.

Shackleton MR.3, WR976 'M', 201 Squadron, RAF Kinloss, 1965 | WR976, a Phase III machine as seen in 1965 (a few months prior to the 'pooling' of the squadrons to form the Kinloss Wing) and has yet to be fitted with Viper jet engines which it received in 1966. WR976 features the repositioned fuselage roundel and 'ROYAL AIR FORCE' legend above the fuselage serial number, Union Flag on the nose and red tip tanks. Finished in the overall gloss Dark Sea Grey scheme it has a white upper fuselage and a black anti-glare panel in front of the windscreen. WR976 was tragically lost, together with nine of the eleven crew on board, when it hit the sea and exploded during a maritime exercise near the Scilly Isles on 19 November 1967.

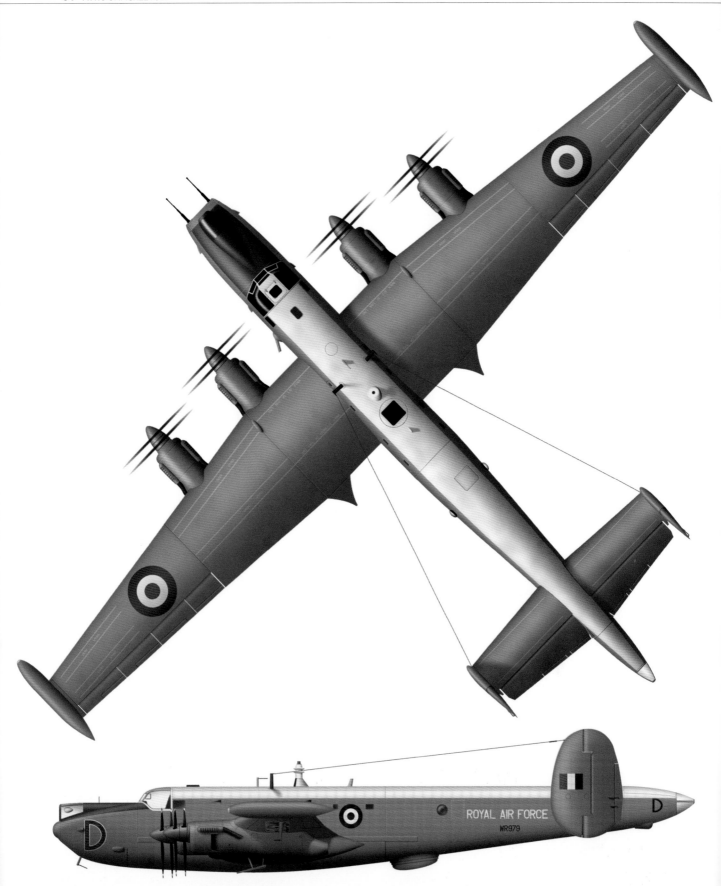

Shackleton MR.3, WR979 'D', of the Kinloss Wing, RAF Kinloss, 1970 | Although squadron numerals still remained evident for some time afterwards, once Shackletons were pooled within the Wing organisation, only the aircraft's *individual* identity letter was required to be displayed, as illustrated here by WR979, coded 'D', which operated with the Kinloss Wing. This letter was repeated, in Dayglo orange, on the rear fuselage immediately forward of the tail cone on Kinloss Wing aircraft, while the St Mawgan Wing applied their tail cone letters in white. A Phase III MR.3, WR979 is illustrated with Viper jet engines and distinctive air intakes (scoops) fitted. Finished in the overall gloss Dark Sea Grey scheme with white upper fuselage, it features the repositioned fuselage roundel and 'ROYAL AIR FORCE' legend above the red fuselage serial number. WR979 remained with the Kinloss Wing until the end of July 1970 when it was flown to St Athan and stored. It was SOC in October 1971 and scrapped soon afterwards.

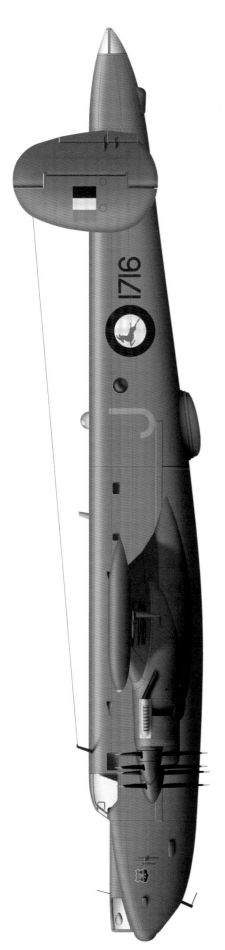

Shackleton MR.3, 1716 'J', 35 Squadron SAAF, Ysterplaat AFB, near Cape Town, late 1950s | Built to the same basic standard as the RAF's MR.3s, eight SAAF Shackletons were operated by 35 Squadron originally for patrolling the sea lanes around the Cape of Good Hope, monitoring Soviet vessels traversing between the Indian and Atlantic oceans. Finished in Dark Sea Grey upper surfaces and PRU Blue under surfaces with the demarcation between the two level with the bottom edge of the main cockpit canopy, the first three Shackletons delivered, 1716 (illustrated here), 1717 and 1719, carried a SAAF roundel with an orange springbok at its centre in all six positions, with an orange, white and blue fin flash on the outer faces of the fins. The serial number was applied in black on both sides of the rear fuselage and under the wings inboard of the national marking, with a yellow individual aircraft letter, 'J', forward of the roundel.

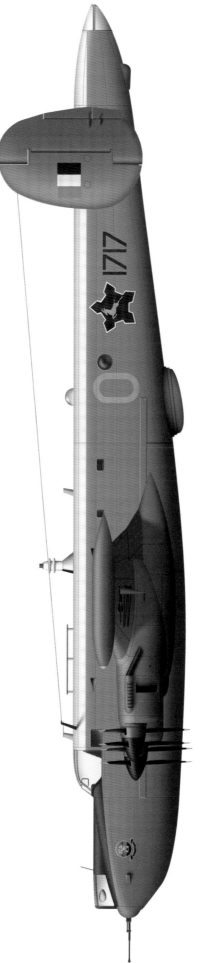

Avro Shackleton MR.3, 1717 'O', 35 Squadron SAAF, D. F. Malan Airport, near Cape Town, 1970s | Because of their weight, SAAF Shackletons had to operate from D. F. Malan Airport as Ysterplaat's runway was not long enough. This meant that crews had to be bussed between their HQ at Ysterplaat and Malan every morning and afternoon. By the mid-late 1960s, all SAAF Shackletons had been marked with the revised SAAF 'castle' insignia, again with an orange springbok at its centre, but still retained the RAF-style fin flash on the outer faces of the fins. Around this time, the original delivery scheme was modified by the addition of a white upper fuselage and a black anti-glare panel to the nose in a similar manner to RAF machines. The addition of the white upper surface left a narrow strip of Dark Sea Grey running the length of the fuselage, but this was eliminated during 1969 by extending the PRU Blue up the side of the fuselage to meet the bottom edge of the white fuselage top as illustrated by 1717. In addition to these changes, the propeller spinners were painted red. The Squadron badge of a pelican astride a globe showing the African continent with the motto *Haya Amanzi* (Strike the Water) was applied in a shield (later changed to a circular design) on the nose, with the last two characters of the serial number applied between the nose guns in black.

Modelling the Shackleton

1/144 SCALE KITS

Czech Master Resin CMRMM02
1/144 Avro Shackleton MR.2/AEW.2
Full resin kit.

Czech Master Resin CMRMM03
1/144 Avro Shackleton MR.1/T.4
Full resin kit.

Welsh Models WHPJW33
1/144 Avro Shackleton AEW/MR.2
Vacform kit.

1/72 SCALE KITS

The original 1967 FROG release of a Shackleton MR.3 is no longer available, but the mouldings have been re-issued by several other manufacturers over the years, including, Novo, Hasegawa, Eastern Express and Revell.

Eastern Express EA72258
1/72 Avro Shackleton MR.3
Ex-FROG kit moulds with 201 Sqn decal sheet.

Revell 04101
1/72 Avro Shackleton MR.3
Ex-FROG kit moulds with 120 Sqn decal sheet.

Revell (née FROG) 1/72 scale Avro Shackleton kit build by Bill Newton
The FROG Shackleton was one of the 'highlight' kits of 1967 having been announced at the Trade Fairs earlier that year. The initial reviews were very complimentary and, apart from the exceptionally heavy surface rivet detail, it was considered to be an excellent kit, and 'raised the bar', even though it was considered 'expensive' at 17/6d (87½p) for the time!

Initially the kit was also marketed in Japan under the Hasegawa branding, and following the demise of FROG/Rovex, the moulds have continued to re-appear every

Top left: Czech Master Resin 1/144 Avro Shackleton MR.2/AEW.2 CMRMM02 box top.

Top right: Czech Master Resin 1/144 Avro Shackleton MR.1/T.4 CMRMM03 box top.

Upper left: Eastern Express EA72258 1/72 Avro Shackleton MR.3 box top.

Lower and bottom left: The original 1967 and 1974 FROG Shackleton MR.3 boxtops.

so often, under the Revell, Novo, Eastern Express, Chematic, Donetsk and Ark labels.

This particular build was based upon a 1996 Revell boxing and a careful look through the components show them to be still cleanly moulded with few sink marks and little flash. The design details are very much of the 1960s style with somewhat heavy detailing and lots of 'working features', including control surfaces, undercarriage and weapons bay doors.

The first task was to remove all rivet detail. A new scalpel blade was used, held at a shallow angle, running it across the surface rather like a wood plane. This was followed by increasingly finer sanding sticks to reduce any residual rivet detail until it was barely visible. Major and/or relevant panel lines were then re-scribed back in. This work is more easily accomplished before the kit is assembled.

Other areas also needed detailing, improving or generally refining to make the model acceptable by today's standards. The main area for adding detail is the forward fuselage interior, basically comprising the flight deck and forward observer's station, the mainwheel wells and the undercarriage.

Starting with the fuselage interior, the cockpit is very visible under the clear canopy and so a new instrument panel, the raised decking for the pilot and co-pilot, rudder pedals, throttle boxes and trim wheels are a must. Little needs to be added aft of the flight deck as the only detail visible is the seating immediately behind the bulkhead. The front observer's station needs a seat adding and the redundant gun sight mount, although very little can be seen in this area.

Once all the interior detail has been added, the fuselage halves can be joined and the join line given any filling and sanding necessary. The join line along the top of the fuselage is not too good having a sort of rippled finish where the mouldings have suffered slightly over the years, however, additional detail added to the top, in the form of the escape hatches and aerial mounts, applied from 5 thou card and other small pieces of sprue, help to create a smoother surface.

The mainwheel wells need to be blanked off fore and aft and some rib detail added to at least give an impression of the structure.

The nacelles were assembled with just the top half of the undercarriage location retained on the two inboard assembles. This gives a good, positive location for the modified undercarriage legs and is in the correct position for some additionally detailed retraction units which were then glued into position. Care must be taken with the construction of the outer nacelles as there are no positive location points.

Adding the front halves of the engine cowlings to the nacelles is quite tricky as the fit is rather poor and in hindsight it might have been better to join the front engine cowling parts to the nacelles before attach-

ing the nacelles to the wings, as a degree of work is required to blend the parts together. Removing the exhausts before attaching the cowlings makes life a little easier. The cowling rings, which have really deep sink marks, are best attached with white glue and only need a light sanding to blend them

Above: The current Revell Shackleton MR.3 box top artwork.

Below: Avro Shackleton MR.3 kit built by Bill Newton

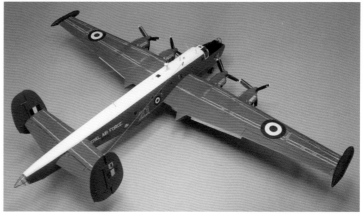

Above and below:
More views of the Avro Shackleton kit built by Bill Newton.

into the rest of the cowling. Being able to remove these parts later allows the rotating propeller assembles to be attached and secured with the retaining bush.

Once all the nacelles are in position and any gaps filled, both wings can then be given a couple of light coats of primer to highlight any further areas that need filling or sanding down and to tone down any remaining visible rivet detail.

The tail units are relatively easy to assemble but again need to be treated to the rivet removal process. [One of the fins in the featured kit had a slight short-shot moulding problem but was easily fixed with card and filler]

Assembly of the wings and tailplanes to the fuselage went well as all the airframe parts fit well without the need for filler. After a final check of the surface detail was made, the assembled airframe was set aside for a few days to allow the filler and primer to harden.

During this time all the smaller ancillary parts can be assembled and painted. The original kit had a so-called 'working under-carriage', (although the undercarriage doors could not be closed), so some modification was necessary to make this area more realistic and to add some of the missing detail. So saying, the top of the main undercarriage legs as moulded were cut away and a new pivot bar installed together with various other details made from plastic sheet, tube and rod to create the retraction jacks and the frame structure. The nosewheel leg also needs a retraction jack. All the various components were then given a coat of white

paint, although the main sections of the legs were painted in a very pale grey to match reference photos. The wheels themselves look relatively accurate so all that was needed was a coat of paint.

Once the undercarriage was fitted, the model needed to be tested for balance on its tricycle under carriage, to avoid a tail-sitting situation. Small lead weights were placed in the nosewheel bay area, until the correct balance was achieved and then the nosewheel side walls, which I had prepared earlier, were added to hide and keep the weights in place.

The propellers need to be cleaned up before being painted. Fortunately the aircraft modelled here had yellow tips to the blades – all twenty-four of them! The later red/white/red propeller tips would have been a bit tedious!

Painting the assembled model was generally straightforward. Three coats, (two matt and the top gloss), of Tamiya White were needed to give the necessary depth of coverage on the top of the fuselage. The area around the exhausts was sprayed with Tamiya Flat Aluminium plus a dash of Gun Metal before being masked off. Then Xtracrylix Dark Sea Grey was applied to the rest of the airframe before finally applying matt black to the anti-glare area forward of the cockpit.

At the time of writing, no aftermarket decals were available for the Shackleton MR.3 so the decal sheet supplied with the kit was used. This offers alternative markings for Nos 120 and 201 Squadrons, RAF. This is not the same as the original FROG sheet, which featured No 206 Sqn RAF and No 35 Sqn SAAF.

A quick test of items on the sheet that were not going to be used showed they worked fine so I went ahead and the decals settled down well over the now relatively smooth surfaces. The model was finished as XF704/L of 201 Sqn. A coat of semi-gloss varnish sealed everything in before the masking tape was removed and then the final details were added. There are a lot of aerials to be attached and careful study of the real subject is recommended.

It is intended to detail this model's weapons bay at some point in the future – but additional research will be needed to find details of the weapons carriers and the stores themselves; maybe raiding the new

Airfix and/or Revell kits for these details might be an option? – so the doors were temporarily fixed in place with a little white glue for the time being.

Job done, and despite its age, the model really looks the part. It will be interesting to see it alongside the new Airfix Shackleton MR.2 and Revell AEW.2.

Avro Shackleton 1/72 AEW.2 conversion by Len Thomson

The basic kit used in this conversion is the 1/72 scale Revell (originally FROG) kit. This particular model was converted using the excellent Aeroclub conversion set. This involved adding a new lower nose section with the large AN/APS-20F radome, converting the fuselage in to a tailwheel configuration, changing the wingtip shape, and adding new canopy window panels. Aeroclub also supply good white metal replacement exhausts.

Those modellers who have built the Revell/FROG/Novo et al model will know that it is covered in hundreds of oversized rivets, so a lot of sanding was needed to remove them all, following which, the model needed to have all its panel lines scribed back in.

The fit of some of the kit parts leaves something to be desired, especially around the engines – and fair bit of filler is needed.

The tail surfaces were also re-worked, which originally featured moving parts, to make them look a little more authentic.

The model was finished as WR963 of No 8 Squadron RAF, based at Lossiemouth. The model was finished in Humbrol Dark Sea Grey overall, with the decals from Aeroclub, which were printed by Fantasy Print Shop.

Above and below: Views of the Avro Shackleton AEW.2 conversion by Len Thomson.

Above: 1/72 scale Gerald Elliott vacform Shackleton MR.1 made straight out of the box by Tony Wilson, and finished as an aircraft of the Gibraltar-based 224 Squadron finished in the Medium Sea Grey and white scheme.

Below: 1/72 scale original FROG Shackleton made by Adrian Morris, and finished as XF701 'F', as a Phase I aircraft of 120 Squadron, when the unit was based at RAF Kinloss in 1959. The model has a detailed open bomb bay, dropped flaps and open fuselage door, and is finished in the overall gloss Dark Sea Grey scheme with white fuselage.

As a large model, the Shackleton lends itself to being displayed in a diorama setting as evidenced by these two 1/72 scale examples on the opposite page.

Sanger CON751
1/72 Avro Shackleton MR.1
Vacform kit

Both Airfix and Revell have announced new tool releases for the end of 2015..

Airfix AX11004
1/72 Avro Shackleton MR.2

Revell RV4920
1/72 Avro Shackleton AEW.2

Below: Images of the much-awaited Airfix 1/72 Avro Shackleton MR.2 kit. *Images © 2015 Hornby Hobbies Ltd. All rights Reserved. Airfix®, a registered trademark of Hornby Hobbies*

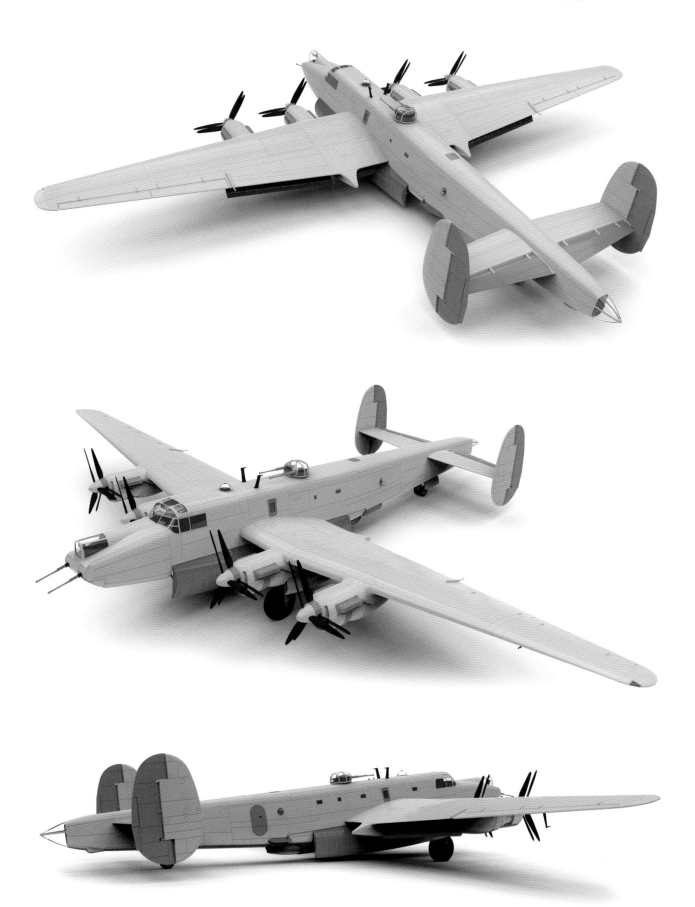

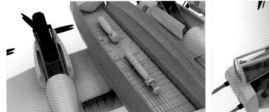

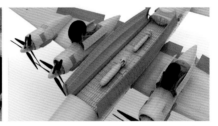

This and opposite page: Further views of the Airfix 1/72 Avro Shackleton MR.2 kit. *Images © 2015 Hornby Hobbies Ltd. All rights Reserved. Airfix®, a registered trademark of Hornby Hobbies*

1/48 SCALE KITS

Sanger SAN4872
1/48 Avro Shackleton MR.1 with decals
Vacform kit

Sanger SAN4873
1/48 Avro Shackleton MR.2 with decals
Vacform kit

Sanger SAN4874
1/48 Avro Shackleton MR.3
Vacform kit

Sanger SAN4876
1/48 Avro Shackleton AEW.2
Vacform kit

ACCESSORIES, DETAILING SETS AND DECALS

Airwaves AEC72198
1/72 Avro Shackleton MR.3 flaps (etched brass) designed to be used with the Eastern Express, FROG, Novo and/or Revell MR.3 kits.

Airwaves AEC72199
1/72 Avro Shackleton MR.3 Interior (etched brass) designed to be used with the Eastern Express, FROG, Novo and/or Revell MR.3 kits.

Czech Master Resin
1/144 Avro Shackleton MR.1/T.4 decals.

Model Alliance ML72210
1/72 RAF Coastal Command Post War Pt 1: featuring Avro Shackleton MR.2 WL737 ToK 220 Sqn St Eval 1953; Avro Lancaster GR.III SW329 JoG 203 Sqn St Eval 1948; Douglas Dakota Mk 4 KN452 Coastal Command Communications Flight Gibraltar 1956; and Short Sunderland GR.V SZ567 230oP 230 Sqn Pembroke Dock 1956.

Model Alliance ML48210
1/48 – Scaled up version of the 1/72 sheet above.

Print Scale PSL72130
1/72 Avro Shackleton MR.3 (Phase 2) XF707, 'C/206' of 206 Sqn, RAF Coastal Command, July 1963 to February 1965; Avro Shackleton MR.1A WB818 of 269 Squadron, with dark blue spinners and squadron badge on nose; Avro Shackleton MR.1A WB859 of 240 Squadron, Ballykelly, 1957.

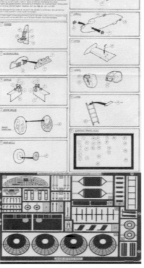

Far left and left: Airwaves accessories.

Below, left to right: Czech Master Resin 1/144 decals for the Avro Shackleton MR.1/T.4 kit.

Model Alliance 1/72 and 1/48 scale decals.

Print Scale 1/72 scale decals.

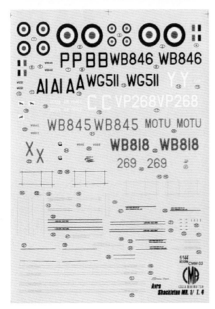

Below and overleaf: Shackleton MR.2 interior detail shots.